COMPUTATIONAL MATHEMATICS AND ANALYSIS

COMPUTATIONAL RECIPES OF LINEAR AND NON-LINEAR SINGULAR INTEGRAL EQUATIONS AND RELATIVISTIC MECHANICS IN ENGINEERING AND APPLIED SCIENCE

VOLUME I

COMPUTATIONAL MATHEMATICS AND ANALYSIS

Additional books in this series can be found on Nova's website under the Series tab.

Additional e-books in this series can be found on Nova's website under the e-book tab.

COMPUTATIONAL MATHEMATICS AND ANALYSIS

COMPUTATIONAL RECIPES OF LINEAR AND NON-LINEAR SINGULAR INTEGRAL EQUATIONS AND RELATIVISTIC MECHANICS IN ENGINEERING AND APPLIED SCIENCE

VOLUME I

EVANGELOS G. LADOPOULOS

New York

Library of Congress Cataloging-in-Publication Data

ISBN: 978-1-63482-450-7
ISSN: 2377-9020

Published by Nova Science Publishers, Inc. † New York

CONTENTS

PREFACE

The present book deals with the computational recipes of the finite-part singular integral equations, the multidimensional singular integral equations and the non-linear singular integral equations, which are widely used in many fields of engineering mechanics and mathematical physics with an applied character, like elasticity, plasticity, thermoelastoplasticity, viscoelasticity, viscoplasticity, fracture mechanics, structural analysis, elastodynamics, fluid mechanics, hydraulics, potential flows and aerodynamics. Such types of linear and non-linear singular integral equations form the latest high technology on the solution of very important problems of solid and fluid mechanics and therefore special attention should be given by the reader of the present book, who is interested for the new computational technology of the twentieth-one century.

Chapter 1 deals with a historical report and an extended outline of References, for the numerical evaluation methods for the finite-part singular integral equations, the multidimensional singular integral equations and the non-linear singular integral equations. Chapter 2 is devoted with the computational recipes for the solution of the finite-part singular integral equations defined in Banach spaces and in general Hilbert spaces. In the same Chapter are proposed and investigated all possible approximation methods for the numerical evaluation of the finite-part singular integral equations, as closed form solutions for the above types of integral equations are available only in simple cases. Also, Chapter 2 provides further several computational integration rules for the solution of the finite-part singular integral equations.

Furthermore, Chapter 3 deals with the application of the finite-part singular integral equations in fracture mechanics and elasticity, by calculating the stress intensity factors in several crack problems which are reduced to the solution of such a type (or systems) of integral equations. Chapter 4 provides further the application of singular integral equations in aerodynamics, by studying planar airfoils in two-dimensional aerodynamics. In Chapter 5 the Singular Integral Operators Method (S.I.O.M.) is introduced and investigated for the numerical evaluation of the multidimensional singular integral equations. This approximation method in many cases offers important advantages over "domain" type solutions, like finite elements and finite difference, as well as analytical methods such as complex variable methods.

Chapter 6 is devoted with the application of the multidimensional singular integral equations in elasticity, viscoelasticity and fracture mechanics of isotropic solids, by considering several two- and three-dimensional elastic stress analysis methods and crack problems. On the other hand, in Chapter 7 is being investigated a special field of applied mechanics, named as Relativistic Mechanics, which is a combination of the classical theory of elasticity and general relativity. Relativistic Mechanics has two main branches Relativistic Elasticity and Relativistic Thermo-Elasticity and according to the above theory, the relative stress tensor for moving structures has been formulated and a formula has been given between the relative stress tensor and the absolute stress tensor of the stationary frame. This leads to the Universal Equation of Elasticity and the Universal Equation of Thermo-Elasticity.

Beyond the above, Chapter 8 deals with the application of the multidimensional singular integral equations in elasticity and fracture mechanics of anisotropic solids, by considering two- and three-dimensional elastic stress analysis. In this case the fundamental solutions of anisotropic stress field analysis are being investigated. Also, Chapter 9 provides further the application of the multidimensional singular integral equations in plasticity of isotropic solids by proposing and studying several applications of two- and three-dimensional plasticity and thermoelastoplasticity.

E. G. Ladopoulos
Civil Engineer, Mechanical Engineer
Interpaper Research Organization
Athens, Greece
E-mail: eladopoulos@interpaper.org

Chapter 1

INTRODUCTION

1.1. NUMERICAL EVALUATION METHODS FOR FINITE-PART SINGULAR INTEGRAL EQUATIONS

A big variety of high technology problems of applied character from elasticity, plasticity, fracture mechanics and aerodynamics are reduced to the solution of finite-part singular integral equations, or to systems of such integral equations. There exists therefore an increasing interest for the numerical evaluation of such systems of singular integral equations of the respective boundary value problems, widely applicable in problems of engineering mechanics.

> The general property of the finite-part singular integral equations, consists to the generalization of the Cauchy singular integral equations, which have been very much investigated over the last decades. The concept of finite-part integrals was firstly introduced by J. Hadamard [1], [2], and L. Schwartz [3] studied some basic properties of them. Some years later, H.R. Kutt [4] proposed some algorithms for the numerical evaluation of the finite-part singular integrals and systematically explained the difference between a finite-part integral and a "generalized principle value integral".

The convergence of several computational methods for the numerical evaluation of finite-part singular integrals was investigated a few years later by M.A. Golberg [5]. He proposed a method, which was an extension beyond the Galerkin and collocation methods [6]. Also, A.C. Kaya and F. Erdogan [7], [8] investigated complicated problems of elasticity and fracture mechanics theory, which are reduced to the solution of finite-part singular integral equations.

Beyond the above, E.G. Ladopoulos [9]-[13] introduced and investigated several computational recipes for the numerical evaluation of the finite-part singular integral equations of the first and the second kind and of systems of such integral equations. He further applied this type of singular integral equations to the solution of several important problems of elasticity, fracture mechanics and aerodynamics. By the same author [14] was investigated a generalization of the Sokhotski-Plemelj formulae valid for the Cauchy-type singular integral equations, in order to show the behaviour of the limiting values of the finite-part singular integrals as well.

Furthermore, E.G. Ladopoulos, V.A. Zisis and D. Kravvaritis [15], [16] used functional analysis as a tool of investigation for the finite-part singular integrals. They studied finite-part singular integral equations defined in general Hilbert spaces and Lp spaces and applied them to several basic crack problems. Also, E.G. Ladopoulos, G. Tsamasphyros and V.A. Zisis [17] investigated computational recipes of finite-part singular integral equations defined in Hilbert spaces, and E.G. Ladopoulos and G. Tsamasphyros approximation methods of singular integral equations in Banach spaces [18]

The finite-part singular integral equations which are studied by the present book, consist to the generalization of the Cauchy singular integral equations almost defined at the end of the 19th century. In 1885 A. Harnack [19] was the first scientist who investigated the limiting values of a singular integral, by splitting it into the sum of two potentials and imposing strong limitations on the density and the contour.

The theory of singular integral equations with their integral in the sense of its principal value was originated quite at the same time with the theory of Fredholm equations. Singular integral equations were investigated by D. Hilbert [20], [21] and H. Poincaré [22], by studying two different problems, Hilbert when investigating some boundary value problems of analytical functions and Poincaré while studying the theory of tides. Moreover, J. Plemelj [23] applied the Cauchy singular integral as a mathematical device for solving a boundary value problem of the theory of analytical functions.

F. Nöther [24], established further the general properties of singular integral equations, by proposing the well-known Nöther theorems. Also N.I. Muskhelishvili [25], [26], investigated the Cauchy-type singular integrals as a mathematical device to various fields of pure and applied mathematics. He has systematically improved the singular integral equations applied to the solution of many important problems of engineering mechanics referring to the theory of elasticity. The solution of singular integral equations of the convolution type was further improved by N. Wiener and E. Hopf [27], when F.D. Gakhov [28], [29] introduced and investigated a general theory of Cauchy singular integral equations in connection with the Rieman-Hilbert boundary value problems.

N.P. Vekua [30], with his well-known monograph, was the first to introduce the method of regularization from the right for the solution of Cauchy-type singular integral equations. Furthermore, the general theory of computational recipes for solving singular integral equations was also improved by V.V. Ivanov [31] and W. Pogorzelski [32], by using some basic topics of functional analysis.

Over the last three decades, the singular integral equation methods with applications to several basic fields of engineering mechanics, like elasticity, plasticity, aerodynamics and fracture mechanics have been studied and improved by several scientists. Among them the following will be mentioned: D. Elliot [33], [34], F. Erdogan, G.D. Gupta and T.S. Cook [35], R.P. Gilbert and R. Magnanini [36], D.A. Hills, D.N. Dai, P.A. Kelly and A.M. Korsunsky [37], G.C. Hsiao, P. Kopp and W.L. Wendland [38], [39], A.M. Korsunsky [40], E.G. Ladopoulos et al. [41] - [44], G. Monegato [45], [46], S. Prossdorf [47], R.P. Srivastav [48], A.N. Teong and D.L. Clements [49], E. Venturino [50], [51], P.P. Zabreyko [52] and U. Zastrow [53], [54].

REFERENCES

[1] J. Hadamard, Lectures on Cauchy's Problem in Linear Partial Differential Equations, Yale University Press, Yale (1932).

[2] J. Hadamard, Le Probléme de Cauchy et les Équations aux Derivées Partielles Linéaires Hyperboliques, Hermann, Paris (1932).

[3] L. Schwartz, *Théorie des Distributions*, Hermann, Paris (1966).

[4] H. R. Kutt, The numerical evaluation of principle value integrals, by finite-part integration, *Num. Math.* 24, 205-210 (1975).

[5] M. A. Golberg, The convergence of several algorithms for solving integral equations with finite-part integrals, *J. Int. Eq.* 5, 329-340 (1983).

[6] M. A. Golberg, The numerical solution of Cauchy singular integral equations with constant coefficients, *J. Int. Eq.* 9, 127-151 (1985).

[7] A. C. Kaya and F. Erdogan, On the solution of integral equations with strongly singular kernels, *Q. Appl. Math.* 45, 105-122 (1987).

[8] A. C. Kaya and F. Erdogan, On the solution of integral equations with a generalized Cauchy kernel, *Q. Appl. Math.* 45, 455-469 (1987).

[9] E. G. Ladopoulos, On the numerical solution of the finite-part singular integral equations of the first and the second kind used in fracture mechanics, *Comp. Meth. Appl. Mech. Engng* 65, 253-266 (1987).

[10] E. G. Ladopoulos, On the numerical evaluation of the general type of finite-part singular integrals and integral equations used in fracture mechanics, *J. Engng Fract. Mech.* 31, 315-337 (1988).

[11] E. G. Ladopoulos, The general type of finite-part singular integrals and integral equations with logarithmic singularities used in fracture mechanics, *Acta Mech.* 75, 275-285 (1988).

[12] E. G. Ladopoulos, Finite-part singular integro-differential equations arising in two-dimensional aerodynamics, *Arch. Mech.* 41, 925-936 (1989).

[13] E. G. Ladopoulos, Systems of finite-part singular integral equations in Lp applied to crack problems, *J. Engng Fract. Mech.* 48, 257-266 (1994).

[14] E. G. Ladopoulos, New aspects for the generalization of the Sokhotski-Plemelj formulae for the solution of finite-part singular integrals used in fracture mechanics, *Int. J. Fract.* 54, 317-328 (1992).

[15] E. G. Ladopoulos, V. A. Zisis and D. Kravvaritis, Singular integral equations in Hilbert space applied to crack problems, *Theor. Appl. Fract. Mech.* 9, 271-281 (1988).

[16] E. G. Ladopoulos, D. Kravvaritis and V. A. Zisis, Finite-part singular integral representation analysis in Lp of two-dimensional elasticity problems, *J. Engng Fract. Mech.* 43, 445-454 (1992).

[17] E. G. Ladopoulos, G. Tsamasphyros and V. A. Zisis, Finite-part singular integral approximations in Hilbert spaces, *Int. J. Math. Math. Scien.* 2004, 2787-2793 (2004).

[18] E. G. Ladopoulos and G. Tsamasphyros, Approximations of singular integral equations on Lyapounov contours in Banach spaces, *Comput. Math. Appl.* . 50, 567-573 (2005).

[19] Harnack, Beiträge zur Theorie des Cauchy'schen Integral, Ber. d. k. 6 Ges. Wiss. , *Math. Phys. Ch.* 37, 379-398 (1885).

[20] D. Hilbert, Über eine Anwendung der Integralgleichungen auf ein Problem der Fuktionentheorie, Verhand. 3 I. Math. Kong. , Heidelberg (1904).
[21] D. Hilbert, Grundzüge einer allgemeinen Theorie der linearen Integral-gleichungen, Leipzig (1912).
[22] H. Poincaré, Lecons de Méchanique Céleste, Vol. 3, Ch. 10, Gauthier et Villars, Paris (1910).
[23] J. Plemelj, Ein Ergänzungssatz zur Cauchyschen Integraldarstellung analytischer Funktionen, Randwerte betreffend, *Monat. Math. Phys.* 19, 205-210 (1908).
[24] F. Nöther, Über eine Klasse singulärer Integralgleichungen, *Math. Annal.* 82, 42-63 (1921).
[25] N. I. Muskhelishvili, *Some Basic Problems of the Mathematical Theory of Elasticity*, Noordhoff, Groningen, The Netherlands (1953).
[26] N. I. Muskhelishvili, *Singular Integral Equations,* Noordhoff, Groningen, The Netherlands (1972).
[27] N. Wiener and E. Hopf, Über eine Klasse singulärer Integralgleichungen, Sitz Berl. Akad. Wiss. , 696-706 (1931).
[28] F. D. Gakhov, On the Rieman boundary value problem, *Matem. Sborn.* 2, 673-683 (1937).
[29] F. D. Gakhov, *Boundary Value Problems*, Pergamon Press, New York (1966).
[30] N. P. Vekua, Systems of Singular Integral Equations, Noordhoff, Groningen, The Netherlands (1967).
[31] V. V. Ivanov, The Theory of Approximate Methods and their Application to the Numerical Solution of Singular Integral Equations, Noordhoff, Leyden, The Netherlands (1976).
[32] W. Pogorzelski, Integral Equations and their Applications, Vol. 1, Pergamon and PWN-Polish Scientific Publ. , Warszawa (1966).
[33] D. Elliot, Orthogonal polynomials associated with singular integral equations having a Cauchy kernel, *SIAM J. Math. Anal.* 13, 1041-1052 (1982).
[34] D. Elliot, The classical collocation method for singular integral equations having a Cauchy kernel, *SIAM J. Numer. Anal.* 19, 816-831 (1982).
[35] F. Erdogan, G. D. Gupta and T. S. Cook, Numerical solution of singular integral equations, in Methods of Analysis and Solutions of Crack Problems (Edited by G. C. Sih), Vol. 1, pp. 368-425, Noordhoff, Leyden, The Netherlands (1973).
[36] R. P. Gilbert and R. Magnanini, The boundary integral method for two-dimensional orthotropic materials, *J. Elasticity* 18, 61-82 (1987).
[37] D. A. Hills, D. N. Dai, P. A. Kelly and A. M. Korsunsky, Singular Integral Equations in the Mechanics of Fracture, Kluwer, Dordrecht (1995).
[38] G. C. Hsiao, P. Kopp and W. L. Wendland, A Galerkin collocation method for some integral equations of the first kind, Computing 25, 83-130 (1980).
[39] G. C. Hsiao, P. Kopp and W. L. Wendland, Some applications of a Galerkin-collocation method for integral equations of the first kind, *Math. Meth. Appl. Sci.* 6, 280-325 (1984).
[40] A. M. Korsunsky, Gauss-Chebyshev quadrature formulae for strongly singular integrals, *Q. Appl. Math.* 56, 461-472 (1998).
[41] E. G. Ladopoulos, On the solution of the two-dimensional problem of a plane crack of arbitrary shape in an anisotropic material, *J. Engng Fract. Mech.* 28, 187-195 (1987).

[42] E. G. Ladopoulos, On a new integration rule with the Gegenbauer polynomials for singular integral equations, used in theory of elasticity, *Ing. Archiv* 58, 35-46 (1988).
[43] V. A. Zisis and E. G. Ladopoulos, Singular integral approximations in Hilbert spaces for elastic stress analysis in a circular ring with curvilinear cracks, *Indus. Math.* 39, 113-134 (1989).
[44] E. G. Ladopoulos, Singular Integral Equations, *Linear and Non-linear Theory and its Applications in Science and Engineering*, Springer-Verlag, Berlin, New York (2000).
[45] G. Monegato, On the weights of certain quadratures for the numerical evaluation of Cauchy principal value integrals and their derivatives, *Numer. Math.* 50, 273-281 (1987).
[46] G. Monegato, Numerical evaluation of hypersingular integrals, *J. Comp. Appl. Math.* 50, 9-31 (1994).
[47] S. Prossdorf, On approximate methods for the solution of one-dimensional singular integral equations, *Appl. Anal.* 7, 259-270 (1977).
[48] R. P. Srivastav, Numerical solution of singular integral equations using Gauss type formulas - I. Quadrature and collocation on Chebyshev nodes, *IMA J. Numer. Anal.* 3, 305-318 (1983).
[49] A. N. Teong and D. L. Clements, A boundary integral equation method for the solution of a class of crack problems, *J. Elasticity* 17, 9-21 (1987).
[50] E. Venturino, Error bounds of the Galerkin method for singular integral equations of the second kind, *BIT* 25, 413-419 (1985).
[51] E. Venturino, Recent developments in the numerical solution of singular integral equations, *J. Math. Anal. Applic.* 115, 239-277 (1986).
[52] P. P. Zabreyko, *Integral Equations-A Reference Text*, Noordhoff, Leyden, The Netherlands (1975).
[53] U. Zastrow, Solution of the anisotropic elastostatical boundary value problems by singular integral equations, Acta Mech. 44, 59-71 (1982).
[54] U. Zastrow, Numerical plane stress analysis by integral equations based on the singularity method, Solid Mech. Arch. 10, 113-128 (1985).

1.2. Numerical Evaluation Methods for Multidimensional Singular Integral Equations

Many important fields of engineering mechanics and mathematical physics, like elasticity, plasticity, thermoelastoplasticity, viscoelasticity, viscoplasticity, elastodynamics, structural analysis, fluid mechanics, hydraulics and fracture mechanics theory, are reduced to the solution of multidimensional singular integral equations, or systems of such integral equations. Thus, several problems of applied character are being solved by methods of multidimensional singular integral equations. These singular integral equations are normally evaluated by numerical methods, as closed form solutions are possible to be determined very seldom and only in special cases.

The first scientist who published some important studies on multidimensional singular integral equations was F.G. Tricomi [1], [2] in 1928, as he investigated double singular integrals by proposing the necessary and sufficient conditions for the existence of such types of multidimensional singular integrals. The theory on multidimensional singular integral equations was further extended by G. Giraud [3] - [5] who investigated singular integrals defined over closed Lyapounov contours of any dimension m.

Furthermore, some years later S.G. Mikhlin [6]-[9] proved that a singular multidimensional operator is bounded in the Hilbert space $L_2(E_m)$, where E_m is an Euclidean space of *m* dimension. At the same time, A.P. Calderon and A. Zygmound [10]-[13] investigated multidimensional singular integrals in the Lebesque-Euclidean space *Lp* (E_m), where $1<p<\infty$ and $p\neq 2$ and proved some very important theorems.

On the other hand, V.D. Kupradze [14] used multidimensional singular integral equations for the solution of three - dimensional problems of elasticity and thermoelasticity of applied character. Also, P.P. Zabreyko [15] in his well monograph proved some important theorems on multidimensional singular integral equations.

Over the last three decades the Boundary Element Merthod (B.E.M.), or Boundary Integral Equation Method (B.I.E.M.) has been formulated and extended investigated for the numerical evaluation of the multidimensional singular integral equations in combination with the solution of some important problems of engineering mechanics, like two- and three-dimensional thermoelastoplasticity, viscoelasticity, viscoplasticity, structural analysis and fracture mechanics. The Boundary Element Method has been investigated and applied in several problems of engineering mechanics and mathematical physics by several scientists. Among them the following will be mentioned: J.D. Achenbach, N. Nishimura and J.C. Sung [16], M.H. Aliabadi, D.P. Rooke and D.J Cartwright [17], N.Altiero and D. Sikarskie [18], J. Balas, V. Sladek and J. Sladek [19], P.K. Banerjee [20], J.Batista de Paiva [21], G. Bezine [22], C.A. Brebbia [23] - [25], C.A. Brebbia, J.C.F. Telles and L.C. Wrobel [26], H.D. Bui [27], J.A.M. Carrer and W.J. Mansur [28],[29], A. Chandra and S. Mukherjee [30], H.B. Chen, P. Lu, M.G. Huang and F.W. Williams [31], W.H. Chen and T.C. Chen [32], S Christiansen and E. Hansen [33], T.A. Cruse [34], T.A. Cruse and W. Vanburen [35], R.P. Gilbert, G.C. Hsiao and M. Schneider [36], R.P. Gilbert and R. Magnanini [37], R.P. Gilbert and M. Schneider [38], L.J. Gray, D. Ghosh and T. Kaplan [39], M. Guiggiani, G. Krishnasamy, T.J. Rudolphi and F.J. Rizzo [40], G.A. Hartley and A. Abdel-Akher [41], G.A. Hartley [42], M. Heinlein, S. Mukherjee and O. Richmond [43], U. Heise [44], J.C. Lachat [45], V. Mantic [46], D. Martin and M. Aliabadi [47], A. Mendelson [48], A. Mendelson and L.U. Albers [49], M. Morjaria and S. Mukherjee [50], S. Mukherjee [51], Y.X. Mukherjee, S.Mukherjee, X. Shi and A. Nagarajan [52], Y.X. Mukherjee, K. Shah and S. Mukherjee [53], A.Nagarajan, S. Mukherjee and E. Lutz [54], K.S. Parihar and S. Sowdamini [55], F. Paris and S. de León [56], H. Poon, S. Mukherjee and M.F. Ahmad [57], N.N.V. Prasad, M.H. Aliabadi and D.P. Rooke [58]-[59], Y.F. Rashed and M.H. Aliabadi [60], Y.F. Rashed, M.H. Aliabadi and C.A. Brebbia [61], F.J. Rizzo and D.J. Shippy [62], M.A. Sales and L.J. Gray [63], D. Segond and A. Tafreshi [64], V. Sladek, J. Sladek and M. Tanaka [65], M.A. Sutton, C.H. Liu, J.R. Dickerson and S.R. McNeill [66], J.L. Swedlow and T.A. Cruse [67], M. Tanaka, V. Sladek and J. Sladek [68], A.N.Teong and D.L. Clements [69], S.M. Vogel and F.J. Rizzo [70], J. Weaver [71], N. Zabaras and S. Mukherjee [72] and others.

On the other hand, by E.G. Ladopoulos [73] - [85] was proposed the Singular Integral Operators Method (S.I.O.M.) for the numerical evaluation of the multidimensional singular integral equations in combination with the solution of several important problems of engineering mechanics and applied science. Especially by the Singular Integral Operators Method are solved problems of two- and three-dimensional elastoplasticity of isotropic and anisotropic solids, applied problems of linear viscoelasticity and viscoplasticity, problems of structural analysis and fracture mechanics problems in isotropic and anisotropic materials [86]. Finally, E.G. Ladopoulos and V.A. Zisis proposed and investigated closed form solutions for the two-dimensional singular integral equations [87].

REFERENCES

[1] F.G. Tricomi, Formula d'inversione dell'ordine di due integrazioni doppie con asterisco, *Rend. Accad. Naz. Lincei.* 3, 535-539 (1926).

[2] F.G. Tricomi, Equazioni integrali contenti il valor principale di un integrale doppio, *Math. Zeit.* 27, 87-133 (1928).

[3] G. Giraud, Sur differentes questions relatives aux equations du type elliptique, *Ann. Sci.Ecole Norm. Sup.* 47, 197-266 (1930).

[4] G. Giraud, Equations a integrales principales, etude suivie d'une application, *Ann.Sci.Ecole Norm. Sup.* 51, 251-372 (1934).

[5] G. Giraud, Equations et systems d'equations on figurent des valeurs principales d'integrales, *C. R. Acad. Sci.* 204, 628-630 (1937).

[6] S.G. Mikhlin, Compounding of multidimensional singular integrals, *Vestnik Leningr. un-ta* 2, 24-41 (1955).

[7] S.G. Mikhlin, The theory of multidimensional singular integral equations, *Vestnik Leningr.un-ta* 1, 3-24 (1956).

[8] S.G. Mikhlin, Notes on the solutions of multidimensional singular integral equations, *Dokl. Akad. Nauk SSSR* 131, 1019-1021 (1960).

[9] S.G. Mikhlin, *Multidimensional Singular Integrals and Integral Equations*, Pergamon Press, Oxford (1965).

[10] A. Calderon and A. Zygmound, On the existence of certain singular integrals, *Acta Math.* 88, 85-139 (1952).

[11] A. Calderon and A. Zygmound, On the singular integrals, *Amer. J. Math.* 78, 289-309 (1956).

[12] A. Calderon and A. Zygmound, Singular integral operators and differential equations, *Amer. J. Math.* 79, 901-921 (1957).

[13] A. Calderon, Singular integrals, *Bull. Amer. Math. Soc.* 72, 426-465 (1966).

[14] V.D. Kupradze, *Three-dimensional Problems in the Mathematical Theory of Elasticity and Thermoelasticity*, Nauka, Moscow (1976).

[15] P.P. Zabreyko, *Integral Equations-A Reference Text,* Noordhoff, Leyden, The Netherlands (1975).

[16] J.D. Achenbach, N. Nishimura and J.C. Sung, Crack - tip fields in a viscoplastic material, *Int J. Solid. Struct.* 23, 1035-1052 (1987).

[17] M.H. Aliabadi, D.P. Rooke and D.J. Cartwright, An improved boundary element formulation for calculating stress intensity factors: Application to aerospace structures, *J. Str. Anal.* 22, 203-207 (1987).

[18] N. Altiero and D. Sikarskie, A boundary integral method applied to plates of arbitrary plane form, *Comp. Struct.* 9, 163-168 (1978).

[19] J. Balas, V. Sladek and J. Sladek, The boundary integral equation method for plates resting on a two parameter foundation, *ZAMM* 51, 574-580 (1984).

[20] P.K. Banerjee, *The Boundary Element Methods in Engineering,* McGraw-Hill, Berkshire (1994).

[21] J. Batista de Paiva, Boundary element formulation of building slabs, *Engng Anal. Bound. Elem.* 17, 105-110 (1996).

[22] G. Bezine, Boundary integral equations for plate flexure with arbitrary boundary conditions, *Mech. Res. Comm.* 5, 197-206 (1978).

[23] C.A. Brebbia, *The Boundary Element Method for Engineers,* Pentech Press, London (1978).

[24] C.A. Brebbia, *The Boundary Element Method for Engineers,* 2nd revised ed., Pentech Press, London (1980).

[25] C.A. Brebbia, *Progresses in the Boundary Element Method*, Pentech Press, London (1981).

[26] C.A. Brebbia, J.C.F. Telles and L.C. Wrobel, *Boundary Element Techniques : Theory and Applications in Engineering,* Springer-Verlag, Berlin (1984).

[27] H.D. Bui, Some remarks about the formulation of three-dimensional thermoelastoplastic problems of integral equations, *Int. J. Sol. Struct.* 14, 935-939 (1978).

[28] J.A.M. Carrer and W.J. Mansur, Time-domain BEM analysis for the 2D scalar wave equation: initial conditions contributions to space and time derivatives, *Int. J. Numer. Meth. Engng* 39, 2169-2188 (1996).

[29] J.A.M. Carrer and W.J. Mansur, Stress and velocity in 2D transient elastodynamic analysis by the boundary element method, *Engng Anal. Bound. Elemen.* 23, 233-245 (1999).

[30] A.Chandra and S.Mukherjee, A boundary element formulation for large strain problems of compressible plasticity, *Engng Anal.* 3, 71-78 (1986).

[31] H.B. Chen, P. Lu, M.G. Huang and F.W. Williams, An effective method for finding values on and near boundaries in the elastic BEM, *Comp. Struct.* 69, 421-431 (1998).

[32] W.H. Chen and T.C. Chen, An efficient dual boundary element method technique for a two-dimensional fracture problem with multiple cracks, *Int. J. Numer. Meth. Engng* 38, 1739-1756 (1995).

[33] S. Christiansen and E. Hansen, A direct integral equation method for compounding the hoop stress at holes in plane, isotropic sheets, *J. Elasticity* 5, 1-14 (1975).

[34] T.A. Cruse, Application of the boundary-integral equation method to three-dimensional stress analysis, *Comp. Struct.* 3, 509-527 (1973).

[35] T.A. Cruse and W. Vanburen, Three-dimensional elastic stress analysis of a fracture specimen with an edge crack, *J. Fract. Mech.* 7, 1-15 (1971).

[36] R.P. Gilbert, G.C. Hsiao and M. Schneider, The two-dimensional linear orthotropic plate, *Appl. Anal.* 15, 147-169 (1983).

[37] R.P. Gilbert and R. Magnanini, The boundary integral method for two-dimensional orthotropic materials, *J. Elasticity* 18, 61-82 (1987).

[38] R.P. Gilbert and M. Schneider, The linear anisotropic plate, *J. Compos. Mater.* 15, 71-78 (1981).

[39] L.J. Gray, D. Ghosh and T. Kaplan, Evaluation of the anisotropic Green's function in three dimensional elasticity, *Comput. Mech.* 17, 255-261 (1996).

[40] M. Guiggiani, G. Krishnasamy, T.J. Rudolphi and F.J. Rizzo, A general algorithm for the numerical solution of hypersingular boundary integral equations, *ASME J. Appl. Mech.* 59, 604-614 (1992).

[41] G.A. Hartley and A. Abdel-Akher, Analysis of building frames, *ASCE J. Struct. Engng* 119, 468-483 (1993).

[42] G.A. Hartley, Development of plate bending elements for frame analysis, *Engng Anal. Bound. Elem.* 17, 93-104 (1996).

[43] M. Heinlein, S. Mukherjee and O. Richmond, A boundary element method analysis of temperature fields and stress during solidification, *Acta Mech.* 59, 58-81 (1986).

[44] U. Heise, The spectra of some integral operators for plane elastostatic boundary value problems, *J. Elasticity* 8, 47-49 (1978).

[45] J.C. Lachat, A further development of the boundary integral technique for elastostatics, Ph.D. thesis, Southampton University (1975).

[46] V. Mantic, A new formula for the C-matrix in the Somigliana identity, *J.Elasticity* 33, 191-201 (1993).

[47] D. Martin and M. Aliabadi, A BE hyper-singular formulation for contact problems using non-conforming discretization, *Comp. Struct.* 69, 557-565 (1998).

[48] A. Mendelson, Boundary integral methods in elasticity and plasticity, *Report No. NASA TND-7418* (1973).

[49] A. Mendelson and L.U. Albers, Application of boundary-integral equations to elastoplastic problems, in *Boundary Integral Equation Method: Computational Applications in Applied Mechanics,* (eds: T.A. Cruse and F.J. Rizzo), p.p.47-84, ASME (1975).

[50] M. Morjaria and S.Mukherjee, Numerical analysis of planar, time-dependent inelastic deformation of plates with cracks by the boundary element method, *Int. J. Sol. Struct.* 17, 127-143 (1981).

[51] S. Mukherjee, Corrected boundary integral equation in planar thermoelastoplasticity, *Int. J . Sol. Struct.* 13, 331-335 (1977).

[52] Y.X. Mukherjee, S. Mukherjee, X. Shi and A. Nagarajan, The boundary contour method for three-dimensional linear elasticity with a new quadratic boundary element, *Engng Anal. Bound. Elem.* 20, 35-44 (1997).

[53] Y.X. Mukherjee, K. Shah and S. Mukherjee, Thermoelastic fracture mechanics with regularized hypersingular boundary integral equations, *Engng Anal. Bound. Elem.* 23, 89-96 (1999).

[54] A. Nagarajan, S. Mukherjee and E. Lutz, The boundary contour method for three dimensional linear elasticity, *ASME J. Appl. Mech.* 63, 278-286 (1996).

[55] K.S. Parihar and S. Sowdamini, Stress distribution in a two-dimensional infinite anisotropic medium with collinear cracks, *J. Elasticity* 15, 193-214 (1985).

[56] F. Paris and S. de León, Thin plates by the boundary element method by means of two Poisson equations, *Engng Anal. Bound. Elem.* 17, 111-122 (1996).

[57] H. Poon, S. Mukherjee and M.F. Ahmad, Use of "simple solutions" in regularizing hypersingular boundary integral equations in elastoplasticity, *ASME J. Appl. Mech.* 65, 39-45 (1998).

[58] N.N.V. Prasad, M.H. Aliabadi and D.P. Rooke, The dual boundary element method for thermoelastic crack problems, *Int. J. Fract.* 66, 255-272 (1994).
[59] N.N.V. Prasad, M.H. Aliabadi and D.P. Rooke, The dual boundary element method for transient thermoelastic crack problems, *Int. J. Solids Struct.* 33, 2695-2718 (1996).
[60] Y.F. Rashed and M.H. Aliabadi, Fundamental solutions for thick foundation plates, *Mech. Res. Commun.* 24, 331-340 (1997).
[61] Y.F. Rashed, M.H. Aliabadi and C.A. Brebbia, A boundary element formulation for a Reissner plate on a Pasternak foundation, *Comp. Struct.* 70, 515-532 (1999).
[62] J. Rizzo and D.J. Shippy, A method for stress determination in plane anisotropic elastic bodies, *J. Comp. Mater.* 4, 36-61 (1970).
[63] M.A. Sales and L.J. Gray, Evaluation of the anisotropic Green's function and its derivatives, *Comp. Struct.* 69, 247-254 (1998).
[64] D. Segond and A. Tafreshi, Stress analysis of three-dimensional contact problems using the boundary element method, *Engng Anal. Bound. Elem.* 22, 199-214 (1998).
[65] V. Sladek, J. Sladek and M. Tanaka, Evaluation of 1/r integrals in BEM formulations for 3-D problems using coordinate multitransformations, *Engng Anal. Bound. Elem.* 20, 229-244 (1997).
[66] M.A. Sutton, C.H. Liu, J.R. Dickerson and S.R. McNeill, The two-dimensional boundary integral equation method in elasticity with a consistent boundary formulation, *Engng Anal.* 3, 73-84 (1986).
[67] J.L. Swedlow and T.A. Cruse, Formulation of boundary integral equation for three-dimensional elasto-plastic body, *Int. J. Sol. Struct.* 7, 1673-1683 (1971).
[68] M. Tanaka, V. Sladek and J. Sladek, Regularization techniques applied to boundary element methods, *Appl. Mech. Rev.* 47, 457-499 (1994).
[69] A.N. Teong and D.L. Clements, A boundary integral equation method for the solution of a class of crack problems, *J. Elasticity* 17, 9-21 (1987).
[70] S.M. Vogel and F.J. Rizzo, An integral equation formulation of three-dimensional anisotropic elastostatic boundary value problems, *J. Elasticity* 3, 203-216 (1973).
[71] J. Weaver, Three-dimensional crack analysis, *Int. J. Sol. Struct.* 13, 321-330 (1977).
[72] N. Zabaras and S.M. Mukherjee, An analysis of solidification problems by the boundary element method, *Int. J. Num. Meth. Engng* 24, 1879-1900 (1987).
[73] E.G. Ladopoulos, On the numerical evaluation of the singular integral equations used in two- and three-dimensional plasticity problems, *Mech. Res. Commun.* 14, 263-274 (1987).
[74] E.G. Ladopoulos, Singular integral representation of three-dimensional plasticity fracture problem, *Theor. Appl. Fract. Mech.* 8, 205-211 (1987).
[75] E.G. Ladopoulos, On the numerical solution of the multidimensional singular integrals and integral equations used in the theory of linear viscoelasticity, *Int. J. Math. Math. Scien.* 11, 561-574 (1988).
[76] E.G. Ladopoulos, Relativistic elastic stress analysis for moving frames, *Rev. Roum. Sci. Tech., Méc. Appl.* 36, 195-209 (1991).
[77] E.G. Ladopoulos, Singular integral operators method for two-dimensional plasticity problems, *Comp. Struct.* 33, 859-865 (1989).
[78] E.G. Ladopoulos, Cubature formulas for singular integral approximations used in three-dimensional elasticity, *Rev. Roum. Sci. Tech., Méc. Appl.* 34, 377-389 (1989).

[79] E.G. Ladopoulos, Singular integral operators method for three-dimensional elasto-plastic stress analysis, *Comp. Struct.* 38, 1-8 (1991).
[80] E.G. Ladopoulos, Singular integral operators method for two-dimensional elasto-plastic stress analysis, *Forsch. Ingen.* 57, 152-158 (1991).
[81] E.G. Ladopoulos, Singular integral operators method for anisotropic elastic stress analysis, *Comp. Struct.* 48, 965-973 (1993).
[82] E.G. Ladopoulos, V.A.Zisis and D.Kravvaritis, Multidimensional singular integral equations in Lp applied to three-dimensional thermoelastoplastic stress analysis, *Comp. Struct.* 52, 781-788 (1994).
[83] E.G. Ladopoulos, *Singular Integral Equations, Linear and Non-linear Theory and its Applications in Science and Engineering,* Springer-Verlag, Berlin, New York (2000).
[84] E.G. Ladopoulos, 3-D elastostatics by coupling method of singular integral equations with finite elements, *Engng Anal. Bound. Elem.* 26, 591-596 (2002).
[85] E.G. Ladopoulos, Coupling of singular integral equation methods and finite elements in 2-D elasticity, *Forsch. Ingen.* 69, 11-16 (2004).
[86] E.G. Ladopoulos and V.A. Zisis, Singular integral representation of two-dimensional shear fracture mechanics problem, *Rev. Roum. Sci. Tech., Méc. Appl.* 38, 617-628 (1993).
[87] V.A. Zisis and E.G. Ladopoulos, Two-dimensional singular integral equations exact solutions, *J Comput. Appl. Math.* 31, 227-232 (1990).

1.3. NUMERICAL EVALUATION METHODS FOR NON-LINEAR SINGULAR INTEGRAL EQUATIONS

Recently, an increasing interest has been concentrated to the solution of many important problems of engineering mechanics and mathematical physics, like structural analysis, fluid mechanics and aerodynamics, by using non-linear singular integral equation methods. By using therefore such type of non-linear integral equations, a major field of problems of applied mechanics character is solved. As in the case of general non-linear operators, the principal approach to the study of these equations is based on various kinds of interaction processes, and in each case the basic form of the non-linear operators is naturally taken into account.

The non-linear singular integral equations of the finite-part and the multidimensional type were introduced and investigated by E.G. Ladopoulos [1] - [15] and E.G. Ladopoulos and V.A. Zisis [16] -[18], in order to be applied for the solution of several basic problems of engineering mechanics. Hence many important problems of structural analysis, fluid mechanics, petroleum engineering and aerodynamics are reduced to the solution of non-linear singular integral equations, or systems of such types of equations. The same authors, established some existence and uniqueness theorems for non-linear singular integral equations defined on Banach spaces.

These non-linear singular integral equations were first defined over finite sets of contours and the existence of solutions was studied for two different kinds of such type of non-linear integral equations, the first and the second kind. Beyond the above, the existence of solutions was further established for non-linear singular integral equations over a finite number of arbitrarily ordered arcs, by introducing a Banach space and a special set defined in the space.

Furthermore, E.G. Ladopoulos [3] - [10], by using the field theory of Green, reduced the problem of the unsteady flow of a two-dimensional NACA airfoil to the solution of a non-linear multidimensional singular integral equation. Such nonlinearity is valid because the source and vortex strength distribution are dependent on the history of the vorticity and source distribution on the NACA airfoil surface.

On the other hand, during the past years, several studies were published, investigating non-linear integral equations of simpler form, without any singularities. Among the scientists who studied non-linear integral equations used in applied mechanics, we shall mention the following: J. Andrews and J.M. Ball [19], S.S. Antman [20], [21], S.S. Antman and E.R. Carbone [22], J.M. Ball [23] - [25], H. Brezis [26], P.G. Ciarlet and P. Destuynder [27], P.G. Ciarlet and J. Necas [28], [29], J.E. Dendy [30], Guo Zhong-Heng [31], H. Hattori [32], D. Hoff and J. Smoller [33], W.J. Hrusa [34], R.C. MacCamy [35] - [37], B. Neta [38], [39], R.W. Ogden [40], R.L. Pego [41], M. Slemrod [42] and O.J. Staffans [43].

REFERENCES

[1] E.G. Ladopoulos, Non-linear integro-differential equations used in orthotropic shallow spherical shell analysis, *Mech. Res.Commun.* 18, 111-119 (1991).

[2] E.G. Ladopoulos, Non-linear integro-differential equations in sandwich plates stress analysis, *Mech. Res. Commun.* 21, 95-102 (1994).

[3] E.G. Ladopoulos, Non-linear singular integral computational analysis for unsteady flow problems, *Renew. Energy* 6, 901-906 (1995).

[4] E.G. Ladopoulos, Non-linear singular integral representation for unsteady inviscid flowfields of 2-D airfoils, *Mech. Res. Commun.* 22, 25-34 (1995).

[5] E.G. Ladopoulos, Non-linear singular integral representation analysis for inviscid flowfields of unsteady airfoils, *Int. J. Non-Lin. Mech.* 32, 377-384 (1997).

[6] E.G. Ladopoulos, Collocation approximation methods for non-linear singular integro-differential equations in Banach spaces, *J. Comp. Appl. Math.* 79, 289-297 (1997).

[7] E.G. Ladopoulos, Non-linear multidimensional singular integral equations in 2-dimensional fluid mechanics analysis, *Int. J. Non-Lin. Mech.* 35, 701-708 (2000).

[8] E.G. Ladopoulos, *Singular Integral Equations, Linear and Non-linear Theory and its Applications in Science and Engineering,* Springer-Verlag, Berlin, New York (2000).

[9] E.G. Ladopoulos, Non-linear unsteady flow problems by multidimensional singular integral representation analysis, *Int. J. Math. Math. Scien.* 2003, 3203-3216 (2003).

[10] E.G. Ladopoulos, Non-linear two-dimensional aerodynamics by multidimen-sional singular integral computational analysis, *Forsch. Ingen.* 68, 105-110 (2003).

[11] E.G. Ladopoulos, Non-linear singular integral equations in elastodynamics, by using Hilbert transformations, *Nonlin. Anal., Real World Appl.* 6, 531-536 (2005).

[12] E.G. Ladopoulos, Unsteady inviscid flowfields of 2-D airfoils by non-linear singular integral computational analysis, *Int. J. Nonlin. Mech.* 46, 1022-1026 (2011).

[13] E.G. Ladopoulos, Non-linear singular integral representation for petroleum reservoir engineering, *Acta Mech.* 220, 247-255 (2011).

[14] E.G. Ladopoulos, Petroleum reservoir engineering by non-linear singular integral equations, *Mech. Engng Res.* 1, 2-11 (2011).

[15] E.G. Ladopoulos, Hydrocarbon reserves exploration by real-time expert seismology and non-linear singular integral equations, *Int. J. Oil Gas Coal Tech.* 5, 299-315 (2012).

[16] E.G. Ladopoulos and V.A. Zisis, Non-linear singular integral approximations in Banach spaces, *Nonlin. Anal., Th. Meth. Appl.* 26, 1293-1299 (1996).

[17] E.G. Ladopoulos and V.A. Zisis, Existence and uniqueness for non-linear singular integral equations used in fluid mechanics, *Appl. Math.* 42, 345-367 (1997).

[18] E.G. Ladopoulos and V.A. Zisis, Non-linear finite-part singular integral equations arising in two-dimensional fluid mechanics, *Nonlin. Anal., Th. Meth. Appl.* 42, 277-290 (2000).

[19] J. Andrews and J.M. Ball, Asymptotic behaviour and changes of phase in one-dimensional nonlinear viscoelasticity, *J. Diff. Eqns* 44, 306-341(1982).

[20] S.S. Antman, Ordinary differential equations of nonlinear elasticity I: Foundations of the theories of nonlinearly elastic rods and shells, *Arch. Ration. Mech. Anal.* 61, 307-351 (1976).

[21] S.S. Antman, Ordinary differential equations of nonlinear elasticity II: Existence and regularity for conservative boundary value problems, *Arch. Ration. Mech. Anal.* 61, 352-393 (1976).

[22] S.S. Antman and E.R. Carbone, Shear and necking instabilities in nonlinear elasticity, *J. Elasticity* 7, 125-151 (1977).

[23] J.M. Ball, Convexity conditions and existence theorems in nonlinear elasticity, *Arch. Ration. Mech. Anal.* 63, 337-403 (1977).

[24] J.M. Ball, Discontinuous equilibrium solutions and cavitation in nonlinear elasticity, *Phil. Trans. R. Soc. Lond.* A306, 557-611 (1982).

[25] J.M. Ball, Remarques sur l'existence et la régularité des solutions d'elastostatique non-linéaire, in *Recent Contributions to Nonlinear Partial Differential Equations,* p.p. 50-62, Pitman, Boston (1981).

[26] H. Brezis, Equations et inéquations non lineaires dans les éspaces vectoriels en dualite, *Ann. Inst. Fourier* 18, 115-175 (1968).

[27] P.G. Ciarlet and P. Destuynder, A justification of a nonlinear model in plate theory, *Comp. Meth. Appl. Mech. Engng* 17, 227-258 (1979).

[28] P.G. Ciarlet and J. Necas, Injectivité presque partout, autocontact, et noninterpénétrabilité en élasticité non linéaire tridimensionnelle, *C. R. Akad. Sci. Paris, Sér I* 301, 621-624 (1985).

[29] P.G. Ciarlet and J. Necas, Injectivity and self-contact in non-linear elasticity, *Arch. Ration. Mech. Anal.* 97, 171-188 (1987).

[30] J.E. Dendy, Galerkin's method for some highly nonlinear problems, *SIAM J. Num. Anal.* 14, 327-347 (1977).

[31] Guo Zhong-Heng, The unified theory of variational principles in nonlinear elasticity, *Arch. Mech.* 32, 577-596 (1980).

[32] H. Hattori, Breakdown of smooth solutions in dissipative nonlinear hyperbolic equations, *Q. Appl. Math.* 40, 113-127 (1982).

[33] D. Hoff and J. Smoller, Solutions in the large for certain nonlinear parabolic systems, *Anal. Non Lin.* 2, 213-235 (1985).

[34] W.J. Hrusa, A nonlinear functional differential equation in Banach space with applications to materials with fading memory, *Arch. Ration. Mech. Anal.* 84, 99-137 (1983).

[35] R.C. MacCamy, Nonlinear Volterra equations on a Hilbert space, *J. Diff. Eqns* 16, 373-393 (1974).

[36] R.C. MacCamy, Stability theorems for a class of functional differential equations, *SIAM J. Appl. Math.* 30, 557-576 (1976).

[37] R.C. MacCamy, A model for non-dimensional, nonlinear viscoelasticity, *Q. Appl. Math.* 35, 21-33 (1977).

[38] B. Neta, Finite element approximation of a nonlinear parabolic problem, *Comput. Math. Appl.* 4, 247-255 (1987).

[39] B. Neta, Numerical solution of a nonlinear integro-differential equation, *J. Math. Anal. Appl.* 89, 598-611 (1989).

[40] R.W. Ogden, Principal stress and strain trajectories in nonlinear elastostatics, *Q. Appl. Math.* 44, 255-264 (1986).

[41] R.L. Pego, Phase transitions in one-dimensional nonlinear viscoelasticity: admissibility and stability, *Arch. Ration. Mech. Anal.* 97, 353-394 (1987).

[42] M. Slemrod, Global existence, uniqueness and asymptotic stability of classical smooth solutions in one-dimensional, nonlinear thermoelasticity, *Arch. Ration. Mech. Anal.* 76, 97-133 (1981).

[43] O.J. Staffans, On a nonlinear hyperbolic Volterra equation, *SIAM J. Math. Anal.* 11, 793-812 (1980).

Chapter 2

COMPUTATIONAL RECIPES OF FINITE-PART SINGULAR INTEGRAL EQUATIONS

2.1. INTRODUCTION

Over the last years an increasing interest has been concentrated on the investigation of finite-part singular integral equations applicable in a major field of problems of engineering mechanics and mathematical physics, like elasticity, plasticity, fracture mechanics and aerodynamics. This type of singular integral equations, consists to the generalization of the Cauchy singular integral equations, which have been systematically studied over the last decades.

J. Hadamard [1], [2] was the first scientist to introduce the concept of finite-part integrals and some basic properties were further analyzed by L. Schwartz [3]. A long time later, H.R. Kutt [4] proposed several numerical formulas for the evaluation of finite-part singular integrals. He also explained the difference between a finite-part integral and a "generalized principal value integral".

Some years later, the convergence of several algorithms for solving finite-part singular integrals was studied by M.A. Golberg [5]. The numerical method he proposed was an extension beyond the two well-known Galerkin and collocation methods [6]. Recently, some complicated problems of elasticity and fracture mechanics theory, which are reduced to the solution of finite-part singular integral equations were investigated by A.C. Kaya and F. Erdogan [7], [8].

Furthermore, E.G. Ladopoulos [9] - [13] introduced several numerical methods for the evaluation of the finite-part singular integral equations of the first and the second kind and has applied them to several important problems of elasticity, fracture mechanics and aerodynamics. Beyond the above, by the same author [14] a generalization of the Sokhotski - Plemelj formulae was given, in order to show the behaviour of the limiting values of the finite-part singular integrals.

On the other hand, by E.G. Ladopoulos, V.A. Zisis and D. Kravvaritis [15], [16] was used functional analysis as a tool of investigation. They studied finite-part singular integral equations defined in general Hilbert and *Lp* spaces and applied them to several important problems of fracture mechanics. Also, E.G. Ladopoulos, G. Tsamasphyros and V.A. Zisis [17] proposed computational recipes of finite-part singular integral equations defined in

Hilbert spaces and E.G. Ladopoulos and G. Tsamasphyros [18] investigated numerical methods for singular integral equations on Lyapounov contours defined in Banach spaces.

The finite-part singular integral equations which are studied by this book are defined as a generalization of the Cauchy singular integral equations almost introduced at the end of the 19th century. In 1885 A. Harnack [19] was the first scientist who investigated the limiting values of a singular integral, by splitting it into the sum of two potentials and imposing strong limitations on the density and the contour.

The theory of singular integral equations with the integral in the sense of its principal value was originated almost at the same time with the theory of Fredholm equations. D. Hilbert [20], [21] and H. Poincaré [22] examined singular integral equations while investigating quite different problems: Hilbert when studying some boundary value problems of analytical functions and Poincaré when improving the theory of tides. Furthermore, J. Plemelj [23] employed the Cauchy type singular integral as a mathematical device for solving a characteristic boundary value problem of the theory of analytical functions.

On the other hand, F. Nöther [24] established the general properties of singular integral equations, which are known as Nöther theorems, and had a considerable influence on subsequent investigations. N.I. Muskhelishvili [25], [26] introduced further the Cauchy singular integral as a mathematical device to various fields of pure and applied mathematics and has systematically improved the singular integral equations applied to the solution of many applied problems of the theory of elasticity. N. Wiener and E. Hopf [27] improved the solution of singular integral equations of the convolution type, while F.D. Gakhov [28], [29] introduced a general theory of singular integral equations in connection with Rieman - Hilbert boundary value problems.

With his well-known monograph, N.P. Vekua [30] was the first scientist to introduce the method of regularization from the right for the solution of Cauchy singular integral equations. Beyond the above, V.V. Ivanov [31] and W. Pogorzelski [32] improved a general theory of computational recipes for the solution of Cauchy-type singular integral equations, in connection with some basic topics of functional analysis.

Over the last three decades, several papers have been published by using singular integral equation methods with applications to some basic aspects of engineering mechanics, like elasticity, plasticity, aerodynamics, fluid mechanics and fracture mechanics theory. Among the scientists who studied such methods, we shall mention the following : D. Elliot [33], [34], F. Erdogan, G.D. Gupta and T.S. Cook [35], R.P. Gilbert and R. Magnanini [36], D.A. Hills, D.N. Dai, P.A. Kelly and A.M. Korsunsky [37], G.C. Hsiao, P. Kopp and W.L. Wendland [38], [39], A.M. Korsunsky [40], E.G. Ladopoulos et al. [41] - [44], G. Monegato [45], [46], S. Prossdorf [47], R.P. Srivastav [48], A.N. Teong and D.L. Clements [49], E. Venturino [50], [51], P.P. Zabreyko [52] and U. Zastrow [53], [54].

2.2. Computational Methods for Singular Integral Equations in General Hilbert Spaces

Over the past years the Gauss-Jacobi quadrature formula has been used successfully for the numerical evaluation of the singular integral equations [35].

Hence, some very basic properties for the Gauss-Jacobi approximation rule, for singular integral equations defined in general Hilbert spaces, are proved by the next Theorems [43].

Definition 2.2.1

Let C(A) be a set of functions continuous on A=[-1,1]^2, and $C^{(\xi)}(A) \equiv \left\{B(t,x) \in C(A); B_t^{(\xi)}, B_x^{(\xi)} \in C(A)\right\}$, $\xi \in Z_+$. By $H_{\rho,\mu}^{(\xi)}$ is denoted the set of functions $H(t,x) \in C^{(\xi)}(A)$ such that $B_t^{(\xi)}(B_x^{(\xi)})$ satisfies a Hölder condition with exponent $0 < \rho, \mu \le 1$ with respect to the variable t(x) on [-1,1].

Definition 2.2.2

For $B \in H_{\rho,\mu}^{(\xi)}$ let:

$$H_t(B_t^{(\xi)};\rho) = \sup_{t_1,t_2,x\in[-1,1]} \left|t_1 - t_2\right|^{-\rho} \left|B_t^{(\xi)}(t_1,x) - B_t^{(\xi)}(t_2,x)\right| \tag{2.2.1}$$

and $H_x(B_x^{(\xi)};\mu)$ be defined by the same way. Furthermore, for functions of a single variable $b \in C^{(\xi)}[-1,1]$ the notation $H_\mu^{(\xi)}$ and $H(b^{(\xi)},\mu)$ is obvious.

Definition 2.2.3
Singular integral equation of the second kind is called an equation of the following form:

$$Ah(x) + \frac{B}{\pi}\int_{-1}^{1} \frac{h(t)}{t-x}\,\mathrm{d}t + \int_{-1}^{1} K(x,t)h(t)\,\mathrm{d}t = f(x) \tag{2.2.2}$$

in which x∈(-1,1), A and B are real numbers, $A^2 + B^2 = 1$, $B \neq 0$, K(x,t) and f(x) are known functions, $K(x,t) \in H_{\rho,\mu}^{(\xi)}, f(x) \in H_\rho^{(\xi)}$, $0 < \rho < \mu \le 1$, and h(t) is the unknown function. Furthermore, (2.2.2) is said to be the dominant equation when K(x,t) = 0, and, for A = 0, (2.2.2) is a singular integral equation of the first kind.

Theorem 2.2.1
Consider the singular integral equation of the second kind (2.2.2), be defined in L_2. The above integral equation is equivalent to a Fredholm equation of the second kind, so that every solution of the former is a solution of the latter and vice versa.

Proof. Generally, a solution of the singular integral equation (2.2.2) is defined as following [26], [35]:

$$h(t) = w(t)\varphi(t) \tag{2.2.3}$$

in which the weight function *w(t)* is defined as:

$$w(t) = (1-t)^a (1+t)^b \tag{2.2.4}$$

with:

$$a = \frac{1}{2\pi i}\ln\left(\frac{A-iB}{A+iB}\right) + F, \quad b = -\frac{1}{2\pi i}\ln\left(\frac{A-iB}{A+iB}\right) + G \tag{2.2.5}$$

where F and G are integers.

Furthermore, the index of the singular integral equation (2.2.2) is given by the relation:

$$v = -(a+b) \tag{2.2.6}$$

Computational Recipes of Singular Integrals in Hilbert Spaces

Let the Hilbert spaces $L_{2,w}$ and $L_{2,1/w}$ be given by the relations:

$$\begin{aligned} L_{2,w} &= \left\{ f \Big/ \int_{-1}^{1} w(t)\left|f\right|^{p} \,\mathrm{d}\,t < \infty \right\} \\ L_{2,1/w} &= \left\{ f \Big/ \int_{-1}^{1} \left[\left|f\right|^{2} / w(x)\right] \mathrm{d}\,x < \infty \right\} \end{aligned} \tag{2.2.7}$$

and assume that:

$$\int_{-1}^{1}\int_{-1}^{1} \left[K^2(x,t)\, w(t)/w(x)\right] \mathrm{d}\,x\,\mathrm{d}\,t < \infty \tag{2.2.8}$$

The Hilbert spaces $L_{2,w}$ and $L_{2,1/w}$ are further defined by the inner products:

$$\begin{aligned} < p,q >_{w} &= \int_{-1}^{1} wpq \,\mathrm{d}\,x, \quad p,q \in L_{2,w} \\ < p,q >_{1/w} &= \int_{-1}^{1} \left[pq/w\right] \mathrm{d}\,x, \quad p,q \in L_{2,1/w} \end{aligned} \tag{2.2.9}$$

and norms:

$$\begin{aligned} \left\|p\right\|_{w}^{2} &= < p,p >, \quad p \in L_{2,w} \\ \left\|p\right\|_{1/w}^{2} &= < p,p >_{1/w}, \quad p \in L_{2,1/w} \end{aligned} \tag{2.2.10}$$

Hence, let the following linear operator $R: L_{2,w} \to L_{2,1/w}$ be defined:

$$R\varphi = Aw(x)\varphi(x) + \frac{B}{\pi}\int_{-1}^{1}\frac{w(t)\varphi(t)}{t-x}\,\mathrm{d}t \tag{2.2.11}$$

and the operator $K: L_{2,w} \to L_{2,1/w}$, which is compact, since $K(x,t) \in H_{\rho,\mu}^{(\xi)}, \xi \in Z_+, 0 < \rho < \mu \le 1$:

$$K\varphi = \int_{-1}^{1} K(x,t)w(t)\varphi(t)\,\mathrm{d}t \tag{2.2.12}$$

By using (2.2.11) and (2.2.12), then the singular integral equation (2.2.2) can be written in the following operator form:

$$R\varphi + K\varphi = f \tag{2.2.13}$$

As operator R is bounded, so its adjoint R^* is defined. Since R is unitary, by multiplying (2.2.13) on the left by R^*, then the Fredholm equation of the second kind is obtained:

$$\varphi + R^* K\varphi = R^* f \tag{2.2.14}$$

which can be further written as:

$$(\Delta\varphi)(x) \equiv \varphi(x) + \int_{-1}^{1} w(t)M(t,x)\varphi(t)\,\mathrm{d}t = D(x) \tag{2.2.15}$$

where:

$$M(t,x) = A\frac{K(t,x)}{w(x)} - \frac{B}{\pi}\int_{-1}^{1}\frac{K(t,\zeta)}{w(\zeta)(\zeta - x)}\,\mathrm{d}\zeta \tag{2.2.16}$$

$$D(x) = A\frac{f(x)}{w(x)} - \frac{B}{\pi}\int_{-1}^{1}\frac{f(\zeta)}{w(\zeta)(\zeta - x)}\,\mathrm{d}\zeta + \frac{c_1}{\omega} \tag{2.2.17}$$

under the condition:

$$c_1 = \int_{-1}^{1} w(t)\varphi(t)\,\mathrm{d}t \tag{2.2.18}$$

with:

$$\omega = \int_{-1}^{1} w(t)\,\mathrm{d}t \tag{2.2.19}$$

which completes the assertion of Theorem 2.2.1.

Theorem 2.2.2

Let by $E_n(\varphi,x)$, $\varphi \in H_{\mu}^{(\xi)}$, $\xi \in Z_+$, $0 < \mu \leq 1$, be defined the error estimate for the Gauss-Jacobi quadrature formula used for the numerical evaluation of a singular integral. Then the following inequality is valid:

$$\left|E_n(\varphi,x)\right| \leq C(a,b,\xi,\mu)H(\varphi^{(\xi)},\mu)\cdot w_{a,b}(x)n^{-\xi-\mu}\ln n \tag{2.2.20}$$

for $x\in(-1,1)$, $C(a,b,\xi,\mu)$ is a positive constant, and for sufficiently large n, $n\in N$, with:

$$w_{a,b}(x) = \begin{cases} w_a(x), & x \in [0,1) \\ w_b(x), & x \in (-1,0] \end{cases} \tag{2.2.21}$$

and:

$$w_a(x) = \begin{cases} 1, & a \geq 1/2 \\ (1-x)^{(2a-1)/4}, & -1/2 < a < 1/2 \\ (1-x)^a, & -1 < a \leq -1/2 \end{cases} \tag{2.2.22}$$

$$w_b(x) = \begin{cases} 1, & b \geq 1/2 \\ (1+x)^{(2b-1)/4}, & -1/2 < b < 1/2 \\ (1+x)^b, & -1 < b \leq -1/2 \end{cases} \tag{2.2.23}$$

Proof. The following Gauss-Jacobi quadrature formula is used for the numerical evaluation of the singular integral in the singular integral equation of the second kind (2.2.2): [35], [43]

$$I_{a,b}(\varphi,x) = \frac{1}{\pi}\int_{-1}^{1} \frac{w(t)\varphi(t)}{t-x}\mathrm{d}t = \sum_{k=1}^{n} C_{k,n}^{(a,b)}(x)\varphi(x_k,n) + E_n(\varphi,x) \tag{2.2.24}$$

where:

$$C_{k,n}^{(a,b)}(x)=\begin{cases}\dfrac{\prod_n^{(a,b)}(x)-\prod_n^{(a,b)}(x_{k,n})}{(x-x_{k,n})p_n^{(a,b)'}(x_{k,n})}, & x\neq x_{k,n}\\[2ex] \dfrac{\prod_n^{(a,b)'}(x_{k,n})}{p_n^{(a,b)'}(x_{k,n})}, & x=x_{k,n}\end{cases} \tag{2.2.25}$$

with $p_n^{(a,b)}$ the Jacobi polynomials of degree n, which are orthogonal on [-1,1] with the weight function $w(x)$.

Furthermore, $\prod_n^{(a,b)}(x)=I_{a,b}(P_n^{(a,b)},x)$, and the algebraic degree of accuracy of the formula is equal to n-1, which completes the assertions of Theorem 2.2.2.

The proof of Theorem 2.2.3, which follows, is obvious.

Theorem 2.2.3.

Consider by $E_n(\varphi,x)$, $\varphi\in C[-1,1]$, $-1<a,\ b<1$, the error function for the Gauss-Jacobi quadrature formula (2.2.24). Then, it is valid:

$$\left\|E_n(\varphi,x)\right\|_{2,1/w}\leq C(a,b)\Psi_{n-1}(\varphi) \tag{2.2.26}$$

where $\Psi_{n-1}(\varphi)$ is the best possible approximation of the function φ by algebraic polynomials of order at most n-1, $n\in N$.

Theorem 2.2.4

Let the system of linear algebraic equations be defined by the relations:

$$\begin{aligned}B\sum_{k=1}^{n}\lambda_{k,n}^{(a,b)}(x_{k,n}-\bar{x}_{i,n-k})^{-1}\cdot z_k&=q_i,\quad i=1,\ldots,n-m\\ \sum_{k=1}^{n}\lambda_{k,n}^{(a,b)}\delta(m)z_k&=c_1\delta(m)\end{aligned} \tag{2.2.27}$$

in which $\delta(m)=1$ for $m=1$ and $\delta(m)=0$ for $m=0$ and $m=-1$, c_1 is a known real number and $\lambda_{k,n}^{(a,b)}$, $k=1,2,\ldots,n$, denote the Christoffel numbers of the Gaussian quadrature formula.

i) For $m=0$ and $m=1$, the system (2.2.27) has the following unique solution:

$$z_k=-B\sum_{i=1}^{n-k}\lambda_{i,n-k}^{(-a,-b)}(\bar{x}_{i,n-k}-x_{k,n})^{-1}q_i+\frac{\delta(m)c_1}{\omega},\quad k=1,\ldots,n \tag{2.2.28}$$

where $\omega = \int_{-1}^{1} w(t)\,\mathrm{d}t$.

ii) For $m = -1$, the system (2.2.27) has the unique solution given by (2.2.28) under the following condition:

$$\sum_{i=1}^{n+1} \lambda_{i,n+1}^{(-a,-b)} q_i = 0 \tag{2.2.29}$$

Proof. For $z = (z_1,\ldots,z_n)$, by using the Lagrange interpolation polynomial, one obtains:

$$(L_{n-1}^{(a,b)} z)(x_{k,n}) = L_{n-1}^{(a,b)}(z, x_{k,n}) = z_k, \quad k = 1,2,\ldots,n \tag{2.2.30}$$

Hence, Theorem 2.2.4 is proved by using the inversion formula for the singular integral equation $(K_0 w L_{n-1}^{(a,b)} z)(x) = L_{n-k}^{(-a,-b)}(q,x)$ with respect to $L_{n-1}^{(a,b)} z$, in which K_0 denotes the dominant part of the singular integral equation of the second kind (2.2.2), defined as:

$$(K_0 h)(x) = A h(x) + \frac{B}{\pi} \int_{-1}^{1} \frac{h(t)}{t - x}\,\mathrm{d}t \tag{2.2.31}$$

Theorem 2.2.5.

Consider the equivalent Fredholm equation of the second kind (2.2.15), uniquely solvable in the Hilbert space $L_{2,w}$ given by (2.2.7), $K \in H_{\rho,\mu}^{(\xi)}, f \in H_{\mu}^{(\xi)}$, $0 < \mu < \rho \leq 1$, and:

$$X(a,\xi) X(K) n^{-\xi-\mu} \|\Delta^1\|_{2,w} < 1 \tag{2.2.32}$$

in which $X(a,\xi)$ denotes a constant determined by the initial data, and:

$$X(K) = \max\left\{ H_t(K_t^{(\xi)}, \rho), H_x(H_x^{(\xi)}, \mu) \right\} \tag{2.2.33}$$

Hence, the following system of linear algebraic equations:

$$z_k - B \sum_{l=1}^{n} \lambda_{i,n}^{(a,b)} \left\{ \sum_{j=1}^{n-1} \lambda_{j,n-1}^{(-a,-b)} (\bar{x}_{j,n-1} - x_{k,n})^{-1} \cdot K(x_{l,n}, \bar{x}_{j,n-1}) \right\} z_l = T_k \tag{2.2.34}$$

where:

$$T_k = -B \sum_{j=1}^{n-1} \lambda_{j,n-1}^{(-a,-b)} (\bar{x}_{j,n-1} - x_{k,n})^{-1} f(\bar{x}_{j,n-1}) + \frac{c_1}{\omega}, \quad k = 1,\ldots,n \tag{2.2.35}$$

has a unique solution $z^* = (z_1^*,\ldots,z_n^*)$ of (2.2.15), converge to an exact solution $\varphi^*(x)$ at the rate:

$$\left\|\varphi^* - \varphi_{n-1}^*\right\|_{2,w} \leq C(a,\xi,\rho,\mu)\theta(f,K)n^{-\xi-\mu}\|\Delta^{-1}\|_{2,w} \tag{2.2.36}$$

in which $C(a,\xi,\rho,\mu)$ denotes a positive constant and:

$$\theta(f,K) = \max\left\{X(K), H(f^{(\xi)},\mu)\right\} \tag{2.2.37}$$

Furthermore:

$$\left|\varphi^*(x) - \varphi_{n-1}^*(x)\right| \leq C(a,\xi,\rho,\mu)n^{-\xi-\mu}\left[w^{-1}(x)L_{n-1}^{(-a,-b)}(x) + w_{-a,-b}(x)\ln n\right] \tag{2.2.38}$$

for $x \in (-1,1)$, where $L_n^{(a,b)}(x)$ is the Lebesque function of the Lagrange interpolation process with nodes $x_{k,n}$, $k = 1,\ldots,n$.

Proof. If the index of the singular integral equation (2.2.2) is $v = 1$, then let us consider the following system of linear algebraic equations:

$$\begin{aligned} \sum_{k=1}^{n} \lambda_{k,n}^{(a,b)}\left[B(x_{k,n} - \bar{x}_{j,n-1})^{-1} + K(x_{k,n}, \bar{x}_{j,n-1})\right]z_k &= f(\bar{x}_{j,n-1}) \\ \sum_{k=1}^{n} \lambda_{k,n}^{(a,b)} z_k &= c_1 \end{aligned} \tag{2.2.39}$$

If Theorem 2.2.4 is used, then the system (2.2.39) is equivalent to the system of linear algebraic equations (2.2.34). Hence, by using Theorems 2.2.2 and 2.2.3, then follows the proof of Theorem 2.2.5.

2.3. Approximation Methods for Systems of Singular Integral Equations in General Hilbert Spaces

The singular integral equations are further replaced by a system of linear algebraic equations for the values of the unknown function at specially chosen points within the range of integration. The proposed method uses a general Hilbert space and establishes the algorithm for finding approximate solutions to a system of singular integral equations. Moreover, the nature of approximation, the rate of convergence to the exact solution and the techniques for error estimation are finally examined.

Let a system of singular integral equations of the first kind be defined as following:

$$\int_{-1}^{1} \frac{h_1(t)}{t-x}\mathrm{d}t + \int_{-1}^{1} K(x,t)h_1(t)\mathrm{d}t + \int_{-1}^{1} M_1(x,t)h_2(t)\mathrm{d}t = f_1(x)$$

$$\int_{-1}^{1} \frac{h_2(t)}{t-x} \mathrm{d}t + \int_{-1}^{1} K(x,t) h_2(t) \mathrm{d}t + \int_{-1}^{1} M_2(x,t) h_1(t) \mathrm{d}t = f_2(x) \tag{2.3.1}$$

in which -1 < x < 1. In eqs. (2.3.1), $h_j(t)$ (j = 1,2) are unknowns, while $K(x,t)$, $M_i(x,t)$ and $f_i(x,t)$ (i = 1,2) are known and satisfy Hölder condition in the closed interval [-1,1].

Eqs. (2.3.1) are solved by the following function [15]:

$$h_i(t) = w(t) u_i(t), \quad i = 1,2 \tag{2.3.2}$$

The weight function $w(t)$ is further given by the relation:

$$w(t) = (1-t)^{1/2+N} (1+t)^{1/2+\Lambda} \tag{2.3.3}$$

where N, Λ are integers such that:

$$-(N+\Lambda) = \mu \tag{2.3.4}$$

is restricted to (±1,0).

Consider the Hilbert spaces $L_{2,w,i}$ and $L_{2,1/w,i}$ (i = 1,2) be given by the formulas:

$$L_{2,w,i} = \left\{ f_i \Big/ \int_{-1}^{1} w(t) |f_i|^2 \mathrm{d}t < \infty \right\}$$

$$L_{2,1/w,i} = \left\{ f_i \Big/ \int_{-1}^{1} \left[|f_i|^2 / w(x) \right] \mathrm{d}x < \infty \right\} \tag{2.3.5}$$

They are valid for:

$$\int_{-1}^{1}\int_{-1}^{1} \frac{\left[K^2(x,t) w(t)\right]}{w(x)} \mathrm{d}x \mathrm{d}t < \infty \tag{2.3.6}$$

Furthermore, the Hilbert spaces $L_{2,w,i}$ and $L_{2,1/w,i}$ have the following inner products:

$$< z,\omega >_w = \int_{-1}^{1} w z \omega \, \mathrm{d}x, \quad z,\omega \in L_{2,w,i}$$

$$< z,\omega >_{1/w} = \int_{-1}^{1} \frac{z\omega}{w} \mathrm{d}x, \quad z,\omega \in L_{2,1/w,i} \tag{2.3.7}$$

and norms:

$$\begin{aligned} \|z\|_w^2 &= (z,z)_w, \quad z \in L_{2,w,i} \\ \|z\|_{1/w}^2 &= (z,z)_{1/w}, \quad z \in L_{2,1/w,i} \end{aligned} \tag{2.3.8}$$

Hence, the bounded linear operator $H: L_{2,w,i} \to L_{2,1/w,i}$ on a Hilbert space can be defined as follows:

$$Hu_i = \int_{-1}^{1} \frac{w(t)u_i(t)}{t-x} \mathrm{d}t \quad (i = 1,2) \tag{2.3.9}$$

Beyond the above, the operators $K: L_{2,w,i} \to L_{2,1/w,i}$, and $M_i: L_{2,w,i} \to L_{2,1/w,i}$ are compact such that following relation exists:

$$Ku_i = \int_{-1}^{1} K(x,t)w(t)u_i(t)\,\mathrm{d}t \tag{2.3.10}$$

and:

$$\begin{aligned} M_1 u_2 &= \int_{-1}^{1} M_1(x,t)w(t)u_2(t)\,\mathrm{d}t \\ M_2 u_1 &= \int_{-1}^{1} M_2(x,t)w(t)u_1(t)\,\mathrm{d}t \end{aligned} \tag{2.3.11}$$

Thus, by making use of eqs. (2.3.9) - (2.3.11) inclusive, eqs. (2.3.1) may be written as:

$$Hu_1 + Ku_1 + M_1u_2 = f_1, \quad Hu_2 + Ku_2 + M_2u_1 = f_2 \tag{2.3.12}$$

As the linear operator H is bounded, then its adjoint H^* is defined.

The existence of solutions for the system of singular integral equations (2.3.1) defined in the general Hilbert space, is proved by the next theorem:

Theorem 2.3.1

Consider the system of singular integral eqs. (2.3.1) or eqs. (2.3.12) in the operator forms defined in the Hilbert spaces $L_{2,w,i}$ and $L_{2,1/w,i}$ $(i = 1,2)$ be given by eqs. (2.3.5). Then, the m-solution of this system of equations is given by:

$$u_{1,m} = \sum_{i=1}^{m} (-1)^i (H^* K)^i (f_1 - M_1 u_2)$$

$$u_{2,m} = \sum_{i=1}^{m} (-1)^i \left[H^* K - H^* M_2 (I + H^* K)^{-1} H^* M_1 \right]^i J \tag{2.3.13}$$

in which:

$$J = H^* f_2 - H^* M_2 (I + H^* K)^{-1} H^* f_1 \tag{2.3.14}$$

Proof. The first of eqs. (2.3.12) can be written as:

$$Hu_1 + Ku_1 = \xi \tag{2.3.15}$$

where:

$$\xi = f_1 - M_1 u_2 \tag{2.3.16}$$

By multiplying eq. (2.3.15) by H^*, then follows the Fredholm equation:

$$u_1 + H^* K u_1 = H^* \xi \tag{2.3.17}$$

Furthermore, application of eq. (2.3.16) results:

$$u_1 = (I + H^* K)^{-1} H^* \xi \tag{2.3.18}$$

in which:

$$\| H^* K \| < 1 \tag{2.3.19}$$

By the use of Neuman's expansion, then the function u_1 can be written as:

$$u_1 = \sum_{m=0}^{\infty} (-1)^m (H^* K)^m \xi \tag{2.3.20}$$

and finally:

$$u_{1,m} = \sum_{i=1}^{m} (-1)^i (H^* K)^i \xi \tag{2.3.21}$$

This proves the expression in the first of eqs. (2.3.13).

Moreover, by multiplying the second of eqs. (2.3.12) by H^*, then follows the Fredholm equation:

$$u_2 + H^* K u_2 + H^* M_2 u_1 = H^* f_2 \tag{2.3.22}$$

A combination of eqs. (2.3.18) and (2.3.22) yields:

$$\left[I + H^* K - H^* M_2 (I + H^* K)^{-1} H^* M_1\right] u_2 = J \tag{2.3.23}$$

where J is given by eq. (2.3.14).

Thus eq. (2.6.23) is reduced to the following form:

$$u_2 = \left[I + H^* K - H^* M_2 (I + H^* K)^{-1} H^* M_1\right]^{-1} J \tag{2.3.24}$$

in which: $\left\| H^* K - H^* M_2 (I + H^* K)^{-1} H^* M_1 \right\| < 1$ (2.3.25)

If the Neuman expansion is applied then follows:

$$u_2 = \sum_{m=0}^{\infty} (-1)^m \left[H^* K - H^* M_2 (I + H^* K)^{-1} H^* M_1\right]^m J \tag{2.3.26}$$

Hence, the second of eqs. (2.3.13) is proved.

Numerical Evaluation

For its numerical evaluation the singular integral equation of the second kind:

$$Ah(x) + \frac{B}{\pi} \int_{-1}^{1} \frac{h(t)}{t - x} \mathrm{d}t + \int_{-1}^{1} K(x,t) h(t) \mathrm{d}t = f(x), \quad -1 < x < 1 \tag{2.3.27}$$

is replaced by a system of linear algebraic equations for the values of the unknown function at certain specially chosen points within the range of integration. In eq. (2.3.27), $h(t)$ is the unknown function while $K(x,t)$ and $f(x)$ are known functions that satisfy the Hölder condition in [-1,1]. The quantities A and B are constants.

Theorem 2.3.2

Consider the characteristic equation corresponding to eq. (2.3.27) be evaluated numerically by following formula:

$$I_n(x) = -\frac{B}{\pi} \sum_{k=0}^{2n} \frac{h_{n,k} \Delta t_k}{x - t_k} + \sum_{k=0}^{2n} \frac{h_{n,k} (A + Bi)(x^{2n+1} + R t_k^{2n+1})}{(2n+1)(x - t_k) x^n t_k^n} \tag{2.3.28}$$

in which:

$$R = \frac{Bi - A}{Bi + A}, \quad \Delta t_k = \frac{2\pi i t_k}{2n+1} \tag{2.3.29}$$

The points $t_k = l^{i\varphi_k}$ (k = 0,1,...,2n) divide the integration interval L = [-1,1] into 2n+1 equal parts. The values of $u_{n,k}$ are given by the relations:

$$h_{n,k} = h_n(k), \quad k = 0,1,\ldots,2n \tag{2.3.30}$$

where:

$$h_n(t) = \frac{1}{2n+1}\sum_{k=0}^{2n} h_{n,k}\frac{t^{2n+1} - t_k^{2n+1}}{(t - t_k)t^n t_k^n} \tag{2.3.31}$$

Proof. Let the characteristic equation corresponding to eq. (2.3.27):

$$Ah(x) + \frac{B}{\pi}\int_L \frac{h(t)}{t-x}\mathrm{d}t = f(x) \tag{2.3.32}$$

Thus, the nth order quadratic interpolation approximation to the function $I(x)$ for the left side of eq. (2.3.32) will be used:

$$I_n(x) = Ah_n(x) + \frac{B}{\pi}\int_L \frac{h_n(t)}{t-x}\mathrm{d}t \tag{2.3.33}$$

in which $h_n(t)$ is given by eq. (2.3.31). The relations:

$$\int_L \frac{t^n}{t-x}\mathrm{d}t = \begin{cases} \pi i t^n, & n = 0,1,\ldots \\ -\pi i t^n, & n = -1,-2,\ldots \end{cases} \tag{2.3.34}$$

are applied to give eq. (2.3.28).

Theorem 2.3.3

Consider $f^{(r)}(t) \in H(A)$ on L and $h(t)$ be the solution of eq. (2.3.32). Then it follows that:

$$|h(t_k) - h_n(t_k)| \le C\frac{\ln n}{n^{r+A}}, \quad k = 0,1,\ldots,2n \tag{2.3.35}$$

in which C is a constant independent of n while $h_{n,k}$ are solutions of:

$$-\frac{B}{\pi}\sum_{k=0}^{2n}\frac{h_{n,k}\Delta t_k}{x_m - t_k} = f(x_m), \quad m = 0,1,\ldots,2n \tag{2.3.36}$$

Here, x_m are the zeros of the polynomial $x_0^{2n+1} + Rt_k^{2n+1}$.

Proof. The solution of the system of linear algebraic eqs. (2.3.36) is given by the relation:

$$h_{n,k} = \frac{\pi}{B}\frac{1}{\Delta t_k} I_k \sum_{m=0}^{2n} I_z \frac{f(x_m)}{t_k - x_m}, \quad k = 0,1,\ldots,2n \tag{2.3.37}$$

where:

$$I_k = \prod_{p=0}^{2n}(x_p - t_k) \Bigg/ \prod_{\substack{p=0 \\ p\neq k}}^{2n}(t_p - t_k)$$

$$I_z = \prod_{p=0}^{2n}(x_m - t_p) \Bigg/ \prod_{p=0}^{2n}(x_m - x_p) \tag{2.3.38}$$

Beyond the above, equations (2.3.38) reduce to the relations:

$$I_k = -\frac{t_k}{2n+1}(1+R), \quad I_z = \frac{x_m}{2n+1}(1+M) \tag{2.3.39}$$

in which:

$$M = \frac{Bi + A}{Bi - A} \tag{2.3.40}$$

Hence, it follows that:

$$h_{n,k} = \frac{B}{\pi}\sum_{m=0}^{2n}\frac{f(x_m)\Delta t_m}{t_k - x_m}, \quad k = 0,1,\ldots,2n \tag{2.3.41}$$

For $f(t)\in H$ on L with A and B being arbitrary where $A^2+B^2 = 1$, then eq. (2.3.32) has the unique solution:

$$h(t) = Af(t) - \frac{B}{\pi}\int_L \frac{f(x)}{x-t}\,\mathrm{d}x \tag{2.3.42}$$

Consider by $\varphi(t)$ the right side of the second of eqs. (2.3.38). The quadratic interpolation approximation of this function is $\varphi_n(t)$. By the same way, $f(t)$ can be approximated as:

$$f_n(x) = -\frac{1}{2n+1}\sum_{m=0}^{2n} f(x_m)\frac{t^{2n+1} - x_m^{2n+1}}{(t - x_m)t^n x_m^n} \tag{2.3.43}$$

Thus by combining eqs. (2.3.34) and (2.3.43), then the result is:

$$\varphi_n(x) = \frac{B}{\pi}\sum_{m=0}^{2n}\frac{f(x_m)\Delta x_m}{t - x_m} + \sum_{m=0}^{2n}\frac{f(x_m)(A - Bi)(t^{2n+1} - t_m^{2n+1})}{(2n+1)(t - x_m)t^n x_m^n} \tag{2.3.44}$$

Further use of eqs. (2.3.41) and (2.3.44) leads to: $h_{n,k} = \varphi_n(t_k), \quad (k = 0,1,...,2n)$.

If the approximation is represented by trigonometric polynomials of order n, then Theorem 2.3.3 is proved.

Theorem 2.3.4

Suppose that eq. (2.3.27) has a unique solution, then the rth derivatives of the second term are in the class $H(a)$ for the range of integration and that the rth derivatives of the regular kernels, with respect to each of the variables, are in the class $H(a)$ on the topological square for the range of integration. Then the absolute value of the difference $\{h(t_k) - h_n(t_k)\}$, where $h(t)$ is the solution of eq. (2.3.27) and $h_n(t_k) = h_{n,k}, \quad (k = 0,1,\ldots,2n)$ form the solution of the following system of linear algebraic equations:

$$-\frac{B}{\pi}\sum_{k=0}^{2n}\left[\frac{1}{x_m - t_k} - \frac{\pi}{B}K(x_m, t_k)\right]h_{n,k}\Delta t_k = f(x_m), \quad m = 0,1,...2n \tag{2.3.45}$$

which satisfy an inequality analogous to eq. (2.3.35).

Proof. It is obvious that eq. (2.3.27) is equivalent to a Fredholm equation of the second kind which has unique solution [26]. As the system of equations (2.3.36) is not singular, then eqs. (2.3.45) are equivalent to a system of linear algebraic equations approximating the Fredholm equations. If these equations have a unique solution, then the order of approximation is equal to that of the exact solution as approximated [55].

Some Special Cases

Let some special cases of eqs. (2.3.32) in which the parameters assume specific values.

Theorem 2.3.5

Suppose that in eq. (2.3.32), $A = \cos \pi a, h(x) = w(x)\varphi(x)$, with $w(x) = (1-x)^a(1+x)^b$ and $a+b$ is an integral number such that a, b >-1.

The points x_k (k = 1,2,...,n+a+b+1) are roots of the Jacobi polynomial $P_{n+a+b+1}^{(a-1,-b)}(x)$, in which $P_n^{(a+1,b)}(t_m)=0 \quad (m=1,2,...,n)$. Equation (2.3.32) can be numerically approximated as:

$$f(x_k) \cong -\frac{\sin \pi a}{\pi}\sum_{m=1}^{n}\frac{C_m \varphi(t_m)}{t_m - x_k} - \frac{\sin \pi a}{\pi}\frac{C_0 \varphi(1)}{1-x_k} \tag{2.3.46}$$

where:

$$C_m = \frac{1}{(1-t_m)P_n^{(a+1,b)'}(t_m)}\int_{-1}^{1} P_n^{(a+1,b)}(t)(1-t)^{a+1}(1+t)^b \frac{1}{t-t_m}\,\mathrm{d}t \tag{2.3.47}$$

and:

$$C_0 = \frac{1}{P_n^{(a+1,b)}(1)}\int_{-1}^{1} P_n^{(a+1,b)}(t)(1-t)^a(1+t)^b\,\mathrm{d}t \tag{2.3.48}$$

Proof. The following formula:

$$\begin{aligned} f(x_k) &\cong -\frac{\sin \pi a}{\pi}\sum_{m=1}^{n}\frac{C_m \varphi(t_m)}{t_m - x} - \frac{\sin \pi a}{\pi}\frac{C_0 \varphi(1)}{1-x} \\ &+ 2^{a+b+1}\sum_{m=1}^{n}\frac{P_{n+a+b+1}^{(-a-1,-b)}(x)\varphi(t_m)}{(1-t_m)P_n^{(a+1,b)'}(t_m)(t_m-x)} - 2^{a+b+1}\frac{P_{n+a+b+1}^{(-a-1,-b)}(x)\varphi(1)}{(1-x)P_n^{(a+1,b)}(1)} \end{aligned} \tag{2.3.49}$$

can be used to evaluate eq. (2.3.32) which is satisfied exactly if $\varphi(x)$ is a polynomial whose order does not exceed (n-1). If the points x_k (k = 1,2,...,n+a+b+1) are roots of the Jacobi polynomial $P_{n+a+b+1}^{(-a-1,-b)}$, then eq. (2.3.49) reduces to eq. (2.3.46).

Lemma 2.3.1

Suppose that $a=-\frac{1}{2}$ and $b=\frac{1}{2}$ for the integration interval (0,1), then eq. (2.3.32) may be evaluated numerically by the formula:

$$\int_0^1 \frac{\varphi(t)}{t-x_k}\sqrt{\frac{t}{1-t}}\,\mathrm{d}t \cong \sum_{m=1}^{n}\frac{C_m \varphi(t_m)}{t_m - x_k}, \quad k=1,2,\ldots,n \tag{2.3.50}$$

where:

$$C_m = \frac{\pi}{n}\sin^2\frac{m\pi}{2n}, \quad m=1,2,\ldots,n-1, \quad C_n = C_0 = \frac{\pi}{2n}$$

$$t_m = \sin^2 \frac{m\pi}{2n}, \quad m = 1,2,\ldots,n, \quad x_k = \sin^2 \frac{2k-1}{4n}\pi, \quad k = 1,2,\ldots,n \tag{2.3.51}$$

Proof. The present lemma results directly from Theorem 2.3.5, because of the following equality:

$$P_n^{(\lambda,\mu)}(\cos\omega) = 2^{-(2n+\lambda+\mu)} \binom{2n+\lambda+\mu}{n} \frac{\cos\left[\left(n + \frac{\lambda+\mu+1}{2}\right)\omega - \left(\lambda + \frac{1}{2}\right)\frac{1}{2}\pi\right]}{(\sin\frac{1}{2}\omega)^{\lambda+1/2}(\cos\frac{1}{2}\omega)^{\mu+1/2}} \tag{2.3.52}$$

where $\lambda = \pm\frac{1}{2}$ and $\mu = \frac{1}{2}$.

2.4. Computational Integration Rules for Finite-Part Singular Integrals

Let the following finite-part singular integral: [11]

$$\Phi(\zeta,\mu) = \Gamma(\mu) \oint_L \frac{w(t)\varphi(t)}{(t-\zeta)^\mu}\,\mathrm{d}t, \quad \mu = 1,2,3,\ldots \tag{2.4.1}$$

in which L denotes the interval $[a,b]$ of the real axis, $w(t)$ is a given weight function defined for every $t \in [a,b]$, $\varphi(t)$ is an analytic function of t in any plane domain S, containing the interval L and $\Gamma(\mu)$ is the Gamma function (where $\Gamma(\mu) = (\mu\text{-}1)!$ for every $\mu = 1, 2, 3\ldots$). Then the complex integral $\Phi(\zeta,\mu)$ is a sectionally analytic function of ζ in the whole plane except L.

Furthermore, if ζ does not belong to L, this type of integral behaves like a regular integral, under the condition that the weight-function $w(t)$ does not present any strong singularities, but only weak singularities at a finite number of points in the interval L.

On the other hand, in the case where the point ζ belongs to L then the integral $\Phi(\zeta)$ diverges and it is usually defined in the sense of the principal value as follows:

$$\Phi(t_0,\mu) = \lim_{\varepsilon\to 0} \Gamma(\mu) \oint_{L-2\varepsilon} \frac{w(t)\varphi(t)}{(t-t_0)^\mu}\,\mathrm{d}t \tag{2.4.2}$$

where $\zeta \equiv t_0 \in L$ and $\mu = 1, 2, 3\ldots$

The integral on the right-hand side of eq. (2.4.2) can be defined on the whole interval L by neglecting a small portion $l \equiv 2\varepsilon$ (where ε is a positive real number), which is cut off by a disk of radius ε, centered at the point $\zeta \equiv t_0$, and the proceeding to the limit as $\varepsilon \to 0$. A sufficient condition for the existence of the integral $\Phi(t_0)$ is that its density function $w(t)\varphi(t)$ be an H-continuous function in the interval L, except at the neighbourhood of its ends, where it may present weak singularities, and that the density function be differentiable.

For the numerical evaluation of the integral (2.4.2), the pole of the integrand at the point $t = \zeta$ has to be taken into account.

Hence, for the case $\mu = 1$, one has:

$$\int_L \frac{w(t)\varphi(t)}{t-\zeta}\mathrm{d}t = \Phi(\zeta) \cong \sum_{K=1}^{n} A_K \frac{\varphi(t_K)}{t_K-\zeta} - 2\varphi(\zeta)\frac{\lambda_n(\zeta)}{J_n(\zeta)} \tag{2.4.3}$$

where the functions $\lambda_n(\zeta)$, $J_n(\zeta)$, are defined as:

$$J_n(\zeta) = \prod_{K=1}^{n}(\zeta - t_K) \tag{2.4.4}$$

$$\lambda_n(\zeta) = -1/2\int_L \frac{w(t)J_n(t)}{t-\zeta}\mathrm{d}t \tag{2.4.5}$$

in which t_K = the abscissae and A_K = the weights.

By differentiating eq. (2.4.3) with respect to ζ, we obtain:

$$\begin{aligned} &⨎_L \frac{w(t)\varphi(t)}{(t-\zeta)^2}\mathrm{d}t \cong \sum_{K=1}^{n} A_K \frac{\varphi(t_K)}{(t_K-\zeta)^2} - 2\frac{\mathrm{d}}{\mathrm{d}\zeta}\left(\varphi(\zeta)\frac{\lambda_n(\zeta)}{J_n(\zeta)}\right) \\ &= \sum_{K=1}^{n} A_K \frac{\varphi(t_K)}{(t_K-\zeta)^2} - 2\varphi'(\zeta)\frac{\lambda_n(\zeta)}{J_n(\zeta)} - 2\varphi(\zeta)\frac{\mathrm{d}}{\mathrm{d}\zeta}\left(\frac{\lambda_n(\zeta)}{J_n(\zeta)}\right) \end{aligned} \tag{2.4.6}$$

Beyond the above, by differentiating eq. (2.4.6) with respect to ζ, results:

$$⨎_L \frac{2w(t)\varphi(t)}{(t-\zeta)^3}\mathrm{d}t \cong \sum_{K=1}^{n} A_K \frac{2\varphi(t_K)}{(t_K-\zeta)^3} - 2\frac{\mathrm{d}^2}{\mathrm{d}\zeta^2}\left[\varphi(\zeta)\frac{\lambda_n(\zeta)}{J_n(\zeta)}\right] \tag{2.4.7}$$

By differentiating by the same way we obtain for a (μ-1) differentiation:

$$(\mu-1)!⨎_L \frac{w(t)\varphi(t)}{(t-\zeta)^{\mu}}\mathrm{d}t \cong \sum_{K=1}^{n} A_K \frac{(\mu-1)!\varphi(t_K)}{(t_K-\zeta)^{\mu}} - 2\frac{\mathrm{d}^{(\mu-1)}}{\mathrm{d}\zeta^{(\mu-1)}}\left[\varphi(\zeta)\frac{\lambda_n(\zeta)}{J_n(\zeta)}\right] \tag{2.4.8}$$

On the other hand as it is valid $\Gamma(\mu) = (\mu-1)! \quad (\mu = 1,2,3,\ldots)$, then follows:

$$⨎_L \frac{w(t)\varphi(t)}{(t-\zeta)^{\mu}}\mathrm{d}t \cong \sum_{K=1}^{n} A_K \frac{\varphi(t_K)}{(t_K-\zeta)^{\mu}} - \frac{2}{(\mu-1)!}\frac{\mathrm{d}^{(\mu-1)}}{\mathrm{d}\zeta^{(\mu-1)}}\left[\varphi(\zeta)\frac{\lambda_n(\zeta)}{J_n(\zeta)}\right]$$

And, also:

$$\Phi(\zeta,\mu) = \Gamma(\mu) \oint_L \frac{w(t)\varphi(t)}{(t-\zeta)^{\mu}} \mathrm{d}t \cong \Gamma(\mu) \sum_{K=1}^{n} A_K \frac{\varphi(t_K)}{(t_K-\zeta)^{\mu}} - 2\frac{\mathrm{d}^{(\mu-1)}}{\mathrm{d}\zeta^{(\mu-1)}}\left[\varphi(\zeta)\frac{\lambda_n(\zeta)}{J_n(\zeta)}\right] \tag{2.4.9}$$

In the special case, where the complex pole ζ of this integral coincides with a point t of L (different from its end-points), then the rules of the numerical integration can be properly extended, in order to be applicable in the case of the principal values of the form (2.4.2).

Then the following quadrature is valid for $\mu = 1$: [11]

$$\text{(i)}\quad \Phi(t_0) \cong \sum_{K=1}^{n} A_K \frac{\varphi(t_K)}{t_K - t_0} - 2\varphi(t_0)K_n(t_0) \tag{2.4.10}$$

for $t \neq t_K$ (K = 1, 2,…, n) and:

$$\text{(ii)}\quad \Phi(t_0) \cong \sum_{\substack{K=1 \\ K\neq m}}^{n} A_K \frac{\varphi(t_K)}{t_K - t_0} + A_m\varphi'(t_0) - 2\varphi(t_0)A_n(t_0) \tag{2.4.11}$$

for $t = t_m \, (m = 1,2,\ldots,n)$

where:

$$K_n(t) = \frac{\lambda_n(t)}{J_n(t)}, \quad t \neq t_K \, (K = 1,2,\ldots,n) \tag{2.4.12}$$

$$A_n(t) = \frac{1}{J_n'(t)}\left[\lambda_n'(t) + \frac{1}{4}A_m\sigma_n''(t)\right], \quad t = t_m \ (m = 1,2,\ldots n) \tag{2.4.13}$$

Furthermore, the numerical calculation of (2.4.10) and (2.4.11) will be estimated for the cases where $\mu = 2,3$ and μ. Also, let us note that in all the above considerations, t_0 is not permitted to coincide with the end points a or b of the interval L. In the special case where the variable t_0 coincides with either a or b the improper integral (2.4.2) must be understood in the finite-part sense [4].

$$\begin{aligned}\text{(i)}\quad \oint_L \frac{w(t)\varphi(t)}{(t-t_0)^2}\mathrm{d}t &\cong \sum_{K=1}^{n} A_K \frac{\varphi(t_K)}{(t_K-t_0)^2} - 2\frac{\mathrm{d}}{\mathrm{d}t_0}\left(\varphi(t_0)K_n(t_0)\right) \\ &= \sum_{K=1}^{n} A_K \frac{\varphi(t_K)}{(t_K-t_0)^2} - 2\varphi'(t_0)K_n(t_0) - 2\varphi(t_0)K_n'(t_0)\end{aligned} \tag{2.4.14}$$

for $t \neq t_K$ (K = 1, 2,…, m).

Hence, by differentiating eq. (2.4.14) one obtains:

$$\oint_L \frac{2w(t)\varphi(t)}{(t-t_0)^3}\mathrm{d}t \cong \sum_{K=1}^{n} A_K \frac{2\varphi(t_K)}{(t_K-t_0)^3} - 2\frac{\mathrm{d}^2}{\mathrm{d}t_0^2}\left(\varphi(t_0)K_n(t_0)\right) \tag{2.4.15}$$

And by a (μ-1) differentiation:

$$(\mu-1)!\oint_L \frac{w(t)\varphi(t)}{(t-t_0)^\mu}\mathrm{d}t \cong \sum_{K=1}^{n} A_K \frac{(\mu-1)!\varphi(t_K)}{(t_K-t_0)^\mu} - 2\frac{\mathrm{d}^{(\mu-1)}}{\mathrm{d}t_0^{(\mu-1)}}\left(\varphi(t_0)K_n(t_0)\right)$$

Finally, we obtain:

$$\Phi(t_0,\mu) = \Gamma(\mu)\oint_L \frac{w(t)\varphi(t)}{(t-t_0)^\mu}\mathrm{d}t \cong \Gamma(\mu)\sum_{K=1}^{n} A_K \frac{\varphi(t_K)}{(t_K-t_0)^2} - 2\frac{\mathrm{d}^{(\mu-1)}}{\mathrm{d}t_0^{(\mu-1)}}\left(\varphi(t_0)K_n(t_0)\right) \tag{2.4.16}$$

for $t \neq t_K (K = 1,2,\ldots n)$

For the second case in the same way by differentiating (2.4.11) we obtain:

$$\oint_L \frac{w(t)\varphi(t)}{(t-t_0)^2}\mathrm{d}t \cong \sum_{\substack{u=1\\u\neq m}}^{n} A_u \frac{\varphi(t_u)}{(t_u-t_0)^3} + A_m\varphi''(t_0) - 2\varphi'(t_0)A_n(t_0) - 2\varphi(t_0)A_n'(t_0) \tag{2.4.17}$$

for $t = t_m (m = 1,2,\ldots n)$

By differentiating again one has:

$$\oint_L \frac{2w(t)\varphi(t)}{(t-t_0)^3}\mathrm{d}t \cong \sum_{\substack{u=1\\u\neq m}}^{n} A_u \frac{2\varphi(t_u)}{(t_u-t_0)^3} + A_m\varphi^{(3)}(t_0) - 2\frac{\mathrm{d}^2}{\mathrm{d}t_0^2}\left(\varphi(t_0)A_n(t_0)\right) \tag{2.4.18}$$

And by a (μ-1) differentiation:

$$(\mu-1)!\oint_L \frac{w(t)\varphi(t)}{(t-t_0)^\mu}\mathrm{d}t \cong \sum_{u=1}^{n} A_u \frac{(\mu-1)!\varphi(t_u)}{(t_u-t_0)^\mu} + A_m\varphi^{(\mu)}(t_0) - 2\frac{\mathrm{d}^{(\mu-1)}}{\mathrm{d}t_0^{(\mu-1)}}\left(\varphi(t_0)A_n(t_0)\right)$$

Finally, we obtain:

$$\Phi(t_0,\mu) = \Gamma(\mu)\oint_L \frac{w(t)\varphi(t)}{(t-t_0)^\mu}\mathrm{d}t \tag{2.4.19}$$

$$\cong \Gamma(\mu)\sum_{u=1}^{n} A_u \frac{\varphi(t_u)}{(t_u-t_0)^\mu} + A_m\varphi^{(\mu)}(t_0) - 2\frac{\mathrm{d}^{(\mu-1)}}{\mathrm{d}t_0^{(\mu-1)}}\left(\varphi(t_0)A_n(t_0)\right) \quad (t = t_m, m = 1,2,\ldots,n)$$

Beyond the above, the functions K_n and A_n will be determined for several integration rules which are encountered in elasticity problems [11].

The Gauss-Legendre Rule

This rule has the weight function: $w(t) = 1$ (in other words, it has no weight function). The functions $\lambda_n(\zeta)$ and $J_n(\zeta)$ for this rule are:

$$J_n(\zeta) = P_n(\zeta)$$
$$\lambda_n(\zeta) = Q_n(\zeta) \tag{2.4.20}$$

in which $P_n(\zeta)$ and $Q_n(\zeta)$ are the Legendre polynomial of degree n and Legendre function of the second kind and order n, respectively. Thus, by substituting relation (2.4.20) into eq. (2.4.12) one obtains:

$$K_n(t) = \frac{\lambda_n(t)}{J_n(t)} = \frac{Q_n(t)}{P_n(t)} \tag{2.4.21}$$

According to the above general results and particularly eqs. (2.4.16) and (2.4.19), it is clear that in the case where $\overline{t_0}$ is selected as a root of the function $Q_n(\overline{t_0}) \equiv 0, P_n(\overline{t_0}) \neq 0$ then $K_n(\overline{t_0}) \equiv 0$ and the principal value (2.4.7) can be approximated as a regular integral, that is:

$$\Phi(\overline{t_0}, \mu) \cong \Gamma(\mu) \sum_{u=1}^{n} A_u \frac{\varphi(t_u)}{(t_u - t_0)^{\mu}} \tag{2.4.22}$$

The set of the points $(\overline{\zeta_{0i}})$ is called to be the set of collocation points for the numerical integration rule under consideration.

Moreover the results for $A_n(t)$ are:

$$A_n(t) = \frac{Q_{n-1}(t)}{P_{n-1}(t)} + \frac{n+1}{2A_m} \frac{1}{1-t^2} = \frac{Q_{n+1}(t)}{P_{n+1}(t)} - \frac{n}{2} A_m \frac{t}{1-t^2} \tag{2.4.23}$$

Applications of this rule are given in Table 2.4.1, for $\mu = 1$.

Table 2.4.1. The Gauss-Legendre numerical integration rule ($w(x)$=1, n=2(1)14)

	$N = 4$			$N = 8$	
0.96780	**21190**		**0.99292**	**12537**	**10609**
0.42970	**74773**	**72613**	**0.86417**	**21023**	**21523**
			0.58650	**08788**	**18819**
			0.20745	**53246**	**99112**
	$N = 6$			$N = 10$	
0.98686	**69984**	**07231**	**0.99558**	**36211**	**14547**
0.75275	**58283**	**49230**	**0.91454**	**95165**	**94484**
0.28059	**49441**	**13258**	**0.73471**	**95809**	**67262**
			0.47529	**42697**	**12834**
			0.16436	**83590**	**13911**

The Lobatto-Legendre rule

Like the Gauss-Legendre rule, this rule has no weight function, $w(t) = 1$, and contains among the abscissae used the end points -1, 1 of the integration interval [-1,1]. The functions (2.4.4) and (2.4.5) are of the form:

$$J_n(\zeta) = P_n(\zeta) - P_{n-2}(\zeta) = \frac{2n-1}{n(n-1)}(\zeta^2 - 1)P'_{n-1}(\zeta) \tag{2.4.24}$$

and:

$$\lambda_n(\zeta) = Q_n(\zeta) - Q_{n-2}(\zeta) = \frac{2n-1}{n(n-1)}(\zeta^2 - 1)Q'_{n-1}(\zeta) \tag{2.4.25}$$

Beyond the above, by taking into account the following relations:

$$\begin{aligned} &J''_n(t)/\sigma'(t) = 0, \quad q'_n(t)/\sigma'_n(t) = Q_{n-1}(t)/P_{n-1}(t) \\ &t = t_m \quad (m = 2,3,\dots,n-1) \quad t \neq t_1, t_n \, (t_1 = 1, \; t_n = -1) \end{aligned} \tag{2.4.26}$$

it can be derived that:

$$A_n(t) = Q'_{n-1}(t)/P'_{n-1}(t) \tag{2.4.27}$$

Applications of this rule are given in Table 2.4.2, for $\mu = 1$.

Table 2.4.2. The Lobatto-Legendre numerical integration rule ($w(x)$=1, n=3(l)15)

	$N = 4$			$N = 10$	
0.62317	**47097**	**82724**	**0.96666**	**25550**	**46562**
			0.80214	**02845**	**23118**
	$N = 6$		**0.52916**	**36689**	**95431**
0.88169	**88950**	**45163**	**0.18470**	**40223**	**21681**
0.34837	**00516**	**19560**			
	$N = 8$				
0.94307	**63694**	**74857**			
0.66899	**39708**	**96066**			
0.24144	**26610**	**09340**			

The Gauss-Chebyshev rule

The weight function of this rule is:

$$w(t) = (1-t)^{\pm 1/2}(1+t)^{\pm 1/2} \tag{2.4.28}$$

Also, the functions $J_n(\zeta)$ and $\lambda_n(\zeta)$ for this case are given by:

$$\begin{aligned} &J_n(\zeta) = T_n(\zeta) \\ &\lambda_n(\zeta) = -\frac{\pi}{2} M_{n-1}(\zeta) \\ &\zeta \in [-1,1] \end{aligned} \tag{2.4.29}$$

in which $T_n(\zeta)$ and $M_n(\zeta)$ denote the Chebyshev polynomials of the first and second kind and degree n. Therefore one has:

$$K_n(t) = -\frac{\pi M_{n-1}(t)}{nT_n(t)} \tag{2.4.30}$$

It is obvious that the set of the collocation points (t_{0i}) coincides with the roots of the Chebyshev polynomials of the second kind:

$$M_{n-1}(t_{0i}) = 0$$

$$t_{0i} = \cos\frac{\pi i}{n}$$

$$(i = 1,2,\ldots,n-1) \tag{2.4.31}$$

Also we have:

$$A_n(t) = -\frac{\pi}{2}\frac{M_{n-2}(t)}{T_{n-1}(t)} + \frac{2n-1}{4}A_m\frac{t}{1-t^2}$$

in the case where $t \equiv t_m$.

The Radau-Legendre Numerical Integration Rule

The weight function $w(t)$ of this rule is as following:

$$w(t) = \frac{1}{\sqrt{t}} \tag{2.4.32}$$

while the integration interval L coincides with [0,1].

So the functions $\lambda_n(\zeta)$ and $J_n(\zeta)$ are:

$$J_n(\zeta) = \frac{(4n-1)}{2n(2n-1)}(1-\zeta)P'_{2n-1}(\sqrt{\zeta}) \tag{2.4.33}$$

and:

$$\lambda_n(\zeta) = \frac{1}{\sqrt{\zeta}}\frac{4n-1}{2n(2n-1)}(1-\zeta)Q'_{2n-1}(\sqrt{\zeta}) \tag{2.4.34}$$

in which P and Q are the Legendre polynomials of the first and the second kind and of degree $2n$-1 respectively. Applications of this rule are given in Table 2.4.3, for $\mu = 1$.

Table 2.4.3. The Radau-Legendre numerical integration rule ($w(x)$=1, n=2(l)11)

	$N=4$			$N=8$	
0.95515	**22921**	**66484**	**0.99183**	**56348**	**66521**
0.23487	**55795**	**28816**	**0.84385**	**31692**	**50343**
-0.74364	**64280**	**74884**	**0.52830**	**56093**	**04982**
			0.10715	**64229**	**00751**
	$N=6$		**-0.33621**	**01079**	**82790**
-0.98396	**67214**	**98314**	**-0.71396**	**77595**	**87774**
0.70076	**35685**	**36436**	**-0.95114**	**40073**	**76222**
0.14724	**83882**	**18905**			
-0.46619	**17362**	**63469**			
-0.90499	**80605**	**41710**			

The Lobatto-Chebyshev Rule

The weight function $w(t)$ of this rule is the same as for the Gauss-Chebyshev rule and contains among the abscissae used the end points ±1 of the integration interval. Let us further consider the case where:

$$w(t) = (1 - t^2)^{-1/2} \tag{2.4.35}$$

then:

$$K_n(t) = \frac{\pi T_{n-1}(t)}{2(1-t^2) M_{n-2}(t)} \tag{2.4.36}$$

and:

$$A_n(t) = -\frac{1}{4} A_m \frac{t}{1-t^2} \tag{2.4.37}$$

The Gauss Numerical Integration Rule with a Logarithmic Singularity

The numerical recipes already outlined in the previous paragraphs can be properly extended in order to include the integrals of the form:

$$I(x) = \oint_0^1 \Gamma(\mu) \ln\left(\frac{1}{t}\right) \frac{\varphi(t)}{(t-x)^{\mu}} \mathrm{d}t \tag{2.4.38}$$

Hence, the weight function is valid as:

$$w(t) = \ln\left(\frac{1}{t}\right)$$

For this case the orthogonal polynomials are given by the recurrence formula:

$$P_0(\zeta) = 1, \quad P_1(\zeta) = \zeta - b_1$$

$$P_n(\zeta)=(\zeta-b_n)P_{n-1}(\zeta)-C_nP_{n-2}(\zeta),\quad n\geq 2 \tag{2.4.39}$$

where the coefficients b_n and C_n are determined. Hence, by using (2.4.5), the functions $\lambda_n(\zeta)$ for the case under consideration are given by the relations:

$$\lambda_n(\zeta)=-\frac{1}{2}\int_0^1 \ln\left(\frac{1}{t}\right)\frac{P_n(t)}{t-\zeta}\mathrm{d}t \tag{2.4.40}$$

Furthermore, because of (2.4.39) we obtain:

$$\lambda_n(\zeta)=(\zeta-b_n)\lambda_{n-1}(\zeta)-C_n\lambda_{n-2}(\zeta),\quad n\geq 2 \tag{2.4.41}$$

For $n = 1$, (2.4.41) takes the form:

$$\lambda_1(\zeta)=(\zeta-b_1)\lambda_0(\zeta)-\frac{1}{2} \tag{2.4.42}$$

In accordance with the definition of the principal value of the integral, one obtains for the function $\lambda_0(\zeta)$ in the interval (0,1):

$$\lambda_0(x)=\frac{1}{2}\lim_{\varepsilon\to 0}\left[\int_0^{x-\varepsilon}\frac{\ln t}{t-x}\mathrm{d}t+\int_{x+\varepsilon}^{1}\frac{\ln t}{t-x}\mathrm{d}t\right]\quad 0<x<1$$

Thus, one has:

$$2\lambda_0(x)=\ln|x|\ln\left|\frac{1-x}{x}\right|-D\frac{(x-1)}{x}+D(1)-\lim_{\varepsilon\to 0}\left[D\left(\frac{\varepsilon}{x}\right)-D\left(-\frac{\varepsilon}{x}\right)\right]\quad 0<x<1 \tag{2.4.43}$$

where:

$$D(\zeta)=-\int_0^{\zeta}\frac{\ln(1-n)}{n}\mathrm{d}n$$

denotes the dilogarithm function.

Moreover, for any value of x other than 0, 1 and t_u (u = 1, 2,…, n) denoting the roots of the polynomials $J_n(\zeta)$:

$$J_n(\zeta)=P_n(\zeta)+d_nP_{n-1}(\zeta),\quad n\geq 1 \tag{2.4.44}$$

in which:

$$d_n = -\frac{P_n(\rho)}{P_{n-1}(\rho)}, \quad \rho = 0 \text{ or } 1 \tag{2.4.45}$$

In the Lobatto method the appropriate polynomials are given by:

$$J_n(\zeta) = P_n(\zeta) + d_n P_{n-1}(\zeta) + e_n P_{n-2}(\zeta), \quad n \geq 2 \tag{2.4.46}$$

where:

$$d_n = -(P_n(0)P_{n-2}(1) - P_n(1)P_{n-2}(0))/D$$

$$e_n = (P_n(0)P_{n-1}(1) - P_n(1)P_{n-1}(0))/D$$

$$D = P_{n-1}(0)P_{n-2}(1) - P_{n-1}(1)P_{n-2}(0)$$

Also, the weights of n-point quadrature rules are:

$$A_u = -2\frac{\lambda_n(t_u)}{J_n'(t_u)}, \quad u = 1,2,\ldots,n \tag{2.4.47}$$

where $\lambda_n(\zeta)$ and $J_n(\zeta)$ denote the functions corresponding to the method employed and t_u are the roots of $J_n(\zeta)$.

Hence, the integral $\int_0^1 \ln\left(\frac{1}{t}\right)\frac{\varphi(t)}{t-\zeta}\mathrm{d}t$ can be approximated by the well-known quadrature formula:

$$I(\zeta) = \int_0^1 \ln\left(\frac{1}{t}\right)\frac{\varphi(t)}{t-\zeta}\mathrm{d}t \cong \sum_{u=1}^{n} A_u \Phi(t_u) - 2\sum_{u=1}^{m}\frac{e_u \lambda_n(\zeta_u)}{J_n(\zeta_u)} \tag{2.4.48}$$

in which: $\Phi(t) = \dfrac{\varphi(t)}{t-\zeta}$

By differentiating eq. (2.4.48) with respect to ζ, we have:

$$\neq\!\!\!\!\!\!\int_0^1 \ln\left(\frac{1}{t}\right)\frac{\varphi(t)}{(t-\zeta)^2}\mathrm{d}t \cong \sum_{u=1}^{n} A_u \frac{\varphi(t_u)}{(t_u-\zeta)^2} - 2\sum_{u=1}^{m} e_u \frac{\mathrm{d}}{\mathrm{d}\zeta}\left[\frac{\lambda_n(\zeta_u)}{J_n(\zeta_u)}\right] \tag{2.4.49}$$

By a (μ-1) differentiation of (2.4.48) results:

$$(\mu-1)!\int_0^1 \ln\left(\frac{1}{t}\right)\frac{\varphi(t)}{(t-\zeta)^{\mu}}\,\mathrm{d}t \cong \sum_{u=1}^{n} A_u \frac{(\mu-1)!\varphi(t_u)}{(t_u-\zeta)^{\mu}} - 2\sum_{u=1}^{m} e_u \frac{\mathrm{d}^{(\mu-1)}}{\mathrm{d}\zeta^{(\mu-1)}}\left[\frac{\lambda_n(\zeta_u)}{J_n(\zeta_u)}\right] \qquad (2.4.50)$$

Finally we have:

$$\begin{aligned}\Phi(\zeta,\mu) = \Gamma(\mu)\int_0^1 \ln\left(\frac{1}{t}\right)\frac{\varphi(t)}{(t-\zeta)^{\mu}}\,\mathrm{d}t &\cong \Gamma(\mu)\sum_{u=1}^{n} A_u \frac{\varphi(t_u)}{(t_u-\zeta)^{\mu}} \\ &- 2\sum_{u=1}^{m} e_u \frac{\mathrm{d}^{(\mu-1)}}{\mathrm{d}\zeta^{(\mu-1)}}\left[\frac{\lambda_n(\zeta_u)}{J_n(\zeta_u)}\right]\end{aligned} \qquad (2.4.51)$$

In eq. (2.4.51) ζ_u are the simple poles of $\Phi(\zeta)$, lying in the interior of a simple-closed curve c surrounding the integration interval [0,1].

Consequently, the roots of the function $\lambda_n(\zeta)$ can be selected as the proper set of collocation points, so that only terms corresponding to poles outside the integration interval remain in the second sum of its right-hand side of (2.7.51).

Applications of this rule are given in Table 2.4.4, for $\mu = 1$.

Table 2.4.4. The Gauss numerical integration rule with a logarithmic singularity $\left(w(x) = \ln(1/x),\ n = 2(1)16\right)$

	N=4			*N*=8	
0.38363	**26889**	**40582 (-2)**	**0.13174**	**02206**	**98576 (-2)**
0.12377	**19864**	**12760**	**0.39524**	**19177**	**18783 (-1)**
0.39448	**00207**	**60224**	**0.13303**	**18039**	**81994**
0.71328	**75345**	**04982**	**0.27233**	**85657**	**58531**
0.94847	**10432**	**33792**	**0.44076**	**95323**	**93504**
			0.61740	**83581**	**59500**
	N=6		**0.77997**	**48923**	**16254**
0.20844	**26070**	**94728 (-2)**	**0.90779**	**26658**	**49610**
0.64464	**28304**	**96628 (-1)**	**0.98448**	**68978**	**97472**
0.21414	**60286**	**94201**			
0.42410	**75094**	**36936**			
0.65160	**44426**	**88373**			
0.84892	**82306**	**31715**			
0.97410	**57929**	**00854**			

The Gauss-Jacobi Numerical Integration Rule

The weight function $w(t)$ of this rule is as following:

$$w(t) = (1-t)^{a}(1+t), \quad (a,b > -1)$$

and the integration interval $L \equiv [-1,1]$. Also the set of orthogonal polynomials $P_n^{(a,\beta)}(t)$ and $Q_n^{(a,\beta)}(t)$ are the Jacobi polynomials of the first and the second kind respectively.

On the other hand, the collocation points (t_{0i}) are defined as roots of the following transcendental equation:

$$\lambda_n^{(a,\beta)}(\zeta) = (\zeta - 1)^a (\zeta + 1)^\beta Q_n^{(a,\beta)}(\zeta) \equiv 0$$

Applications of this rule are given in Table 2.4.5, for $\mu = 1$.

Table 2.4.5. Collocation points, abscissae and weights for the Gaussi-Jacobi rule $w(x) = (1-x)^A (1+x)^B$ A, B= – 0.9 (0.2) – 0.1, n=4(1) 8(2) 10

	$X(K)$ A=-0.1			$T(1)$			$N(1)$ B=–0.9	
				$N=4$				
0.57842	**23679**	**14642**	**0.84494**	**75992**	**80885**	**0.28725**	**45307**	**66244**
-0.16960	**08083**	**32198**	**0.22222**	**04969**	**55233**	**0.65705**	**99709**	**31763**
-0.82495	**95232**	**66215**	**-0.53738**	**65810**	**58362**	**0.12894**	**46881**	**01794(+1)**
			-0.98692	**43723**	**20613**	**0.79326**	**46001**	**91458(+1)**
				$N=6$				
0.80535	**91667**	**92133**	**0.93035**	**70605**	**38274**	**0.13622**	**50438**	**96728**
0.40211	**33365**	**95500**	**0.62521**	**19578**	**35826**	**0.29062**	**51603**	**24674**
-0.11029	**35922**	**78042**	**0.15124**	**79156**	**77541**	**0.46414**	**83043**	**65640**
-0.59459	**02985**	**20716**	**-0.36468**	**72310**	**36701**	**0.71432**	**33952**	**27087**
-0.92084	**62856**	**90698**	**-0.78431**	**44804**	**05640**	**0.12388**	**34286**	**85429(+1)**
			-0.99417	**88589**	**72937**	**0.73222**	**51193**	**96210(+1)**
				$N=8$				
0.88905	**47644**	**76471**	**0.96068**	**15552**	**29432**	**0.80715**	**64949**	**03630(-1)**
0.64715	**20044**	**06231**	**0.78320**	**34785**	**63805**	**0.16855**	**37330**	**39112**
0.30628	**61742**	**79217**	**0.48611**	**79538**	**31080**	**0.25670**	**98035**	**93985**
-0.81675	**01746**	**72416 (-1)**	**0.11456**	**61130**	**12191**	**0.35848**	**54344**	**21786**
-0.45767	**00339**	**11592**	**-0.27489**	**57247**	**59346**	**0.49325**	**18664**	**75814**
0.76444	**65247**	**90089**	**-0.62297**	**24478**	**60676**	**0.70770**	**48587**	**85621**
0.95520	**65456**	**29669**	**-0.87664**	**39088**	**97736**	**0.11856**	**68613**	**71647(+1)**
			-0.99672	**36857**	**85416**	**0.69153**	**16425**	**10737(+1)**

The Lobatto-Jacobi Numerical Integration Rule

For this integration rule the collocation points (t_{0i}) are the roots of the following transcendental equation:

$$\lambda_n^{(a,\beta)}(\zeta) = (\zeta - 1)^a (\zeta + 1)^\beta Q_n^{(a,\beta)}(\zeta) \equiv 0$$

Applications of this rule are given in Table 2.4.6, for $\mu = 1$.

Table 2.4.6. Collocation points, abscissae and weights for the Lobatto-Jacobi rule $w(x) = (1-x)^A (1+x)^B$**, *A*, *B*= – 0.9 (0.2) – 0.1, *n*=4(1) 8(2) 10**

X(*K*)			*T*(1)			*W*(1)		
A =–0.1						*B* =–0.9		
				N = 4				
0.77069	**50866**	**32802**	**0.10000**	**00000**	**00000(+1)**	**0.14207**	**13917**	**63439**
-0.18785	**12879**	**42328**	**0.33355**	**85071**	**70123**	**0.80702**	**47313**	**22645**
-0.94020	**16019**	**72955**	**-0.65355**	**85071**	**70123**	**0.19423**	**88574**	**62052(+1)**
			-0.10000	**00000**	**00000(+1)**	**0.72749**	**22686**	**92392(+1)**
				N = 6				
0.91521	**60818**	**34127**	**0.10000**	**00000**	**00000(+1)**	**0.56694**	**73749**	**80632(-1)**
0.49120	**87398**	**05436**	**0.73911**	**69789**	**41872**	**0.29971**	**71144**	**37000**
-0.11777	**73418**	**75109**	**0.19593**	**03066**	**30473**	**0.53865**	**90092**	**35008**
-0.67885	**37200**	**93961**	**-0.41920**	**11706**	**30897**	**0.89815**	**87926**	**75190**
-0.97851	**60960**	**08183**	**-0.87140**	**16704**	**97003**	**0.18102**	**93110**	**15035(+1)**
			0.10000	**00000**	**00000(+1)**	**0.65628**	**84620**	**63490(+1)**
				N = 8				
0.95642	**53062**	**94898**	**0.10000**	**00000**	**00000(+1)**	**0.30946**	**64848**	**16778(-1)**
0.72800	**68251**	**44303**	**0.86389**	**07372**	**23107**	**0.16058**	**72771**	**21863**
0.35601	**25289**	**98959**	**0.55585**	**00214**	**61376**	**0.27331**	**72926**	**76481**
-0.85715	**47390**	**88598(-1)**	**0.13852**	**13128**	**77907**	**0.40232**	**29128**	**91205**
-0.50967	**45273**	**47895**	**-0.30545**	**32455**	**07834**	**0.57791**	**86685**	**79069**
-0.83191	**92578**	**46593**	**-0.68814**	**25252**	**38306**	**0.88009**	**02749**	**56232**
-0.98304	**52051**	**52118**	**-0.93383**	**70704**	**47020**	**0.17068**	**84572**	**90680(+1)**
			-0.10000	**00000**	**00000(+1)**	**0.61343**	**39737**	**01720(+1)**

The Generalized Gauss-Laguerre Numerical Integration Rule

This numerical rule has the following weight function:

$$w(t) = t^a e^{-t}$$

and integration interval $L \equiv [0,\infty]$.

Beyond the above, the collocation points are defined as roots of the function $\lambda_n(\zeta)$, which are estimated by the following recurrence relations:

$$n\lambda_n(\zeta) = (2n - 1 - \zeta)\lambda_{n-1}(\zeta) - (n-1)\lambda_{n-2}(\zeta)$$

$$\lambda_0(\zeta) = \frac{1}{2} e^{-\zeta} E_1(\zeta)$$

$$\lambda_1(\zeta) = (1-\zeta)\lambda_0(\zeta)$$

$$E_1(\zeta) = \int_{-\infty}^{\zeta} e^t / t \, \mathrm{d}t \tag{2.4.52}$$

Applications for this rule are given in Table 2.4.7, for $\mu = 1$.

Table 2.4.7. Collocation points, abscissae and weights for the Gauss-Laguerre rule $w(x) = x^a \exp(-x)$, **a= – 0.9 (0.2) – 0.7, n=4(1)8, 10**

	X(K)			***T*(1)**			*W*(1)	
				***A* =–0.9**				
				N = 4				
0.35914	**33754**	**55156**	**0.25907**	**77815**	**97432(-1)**	**0.88892**	**37184**	**75298(+1)**
0.20230	**57578**	**73572(+1)**	**0.10163**	**33231**	**92219(+1)**	**0.58904**	**86627**	**56540**
0.53343	**71462**	**95893(+1)**	**0.34308**	**74722**	**70107(+1)**	**0.34879**	**57099**	**89796(-1)**
0.11736	**57649**	**81479(+2)**	**0.79268**	**84267**	**21699(+1)**	**0.34228**	**01602**	**28392(-3)**
				N = 6				
0.23950	**21025**	**97005**	**0.17342**	**74091**	**62196(-1)**	**0.86104**	**58012**	**50877(+1)**
0.13222	**72349**	**21621(+1)**	**0.67256**	**50982**	**20954**	**0.78743**	**14118**	**75148**
0.33271	**18158**	**74294(+1)**	**0.22009**	**00258**	**35410(+1)**	**0.10871**	**22810**	**33294**
0.64462	**63393**	**09349(+1)**	**0.47284**	**81828**	**27537(+1)**	**0.67747**	**22140**	**24804(-2)**
0.11135	**46059**	**16006(+2)**	**0.85451**	**61044**	**14663(+1)**	**0.13090**	**24113**	**82447(-3)**
0.19035	**08722**	**06004(+2)**	**0.14935**	**54903**	**00867(+2)**	**0.36869**	**98889**	**94601(-6)**
				N=8				
0.17975	**69252**	**16368**	**0.13033**	**85728**	**66258(-1)**	**0.84032**	**66392**	**07882(+1)**
0.98580	**25615**	**85796**	**0.50346**	**02097**	**34729**	**0.90214**	**59004**	**59557**
	X(K)			***T*(1)**			*W*(1)	
0.24476	**49029**	**09880(+1)**	**0.16315**	**50144**	**42472(+1)**	**0.18373**	**18641**	**50728**
0.46320	**39749**	**94804(+1)**	**0.34435**	**79433**	**77284(+1)**	**0.22883**	**69627**	**75134(-1)**
0.76606	**22503**	**97354(+1)**	**0.60300**	**34104**	**61902(+1)**	**0.14416**	**43517**	**34506(-2)**
0.11762	**16154**	**03866(+2)**	**0.95558**	**69544**	**25233(+1)**	**0.37902**	**18433**	**16068(-4)**
0.17441	**73706**	**17078(+2)**	**0.14350**	**75027**	**04787(+2)**	**0.29968**	**00686**	**85111(-6)**
0.26478	**89930**	**10109(+2)**	**0.21271**	**72243**	**59311(+2)**	**0.32036**	**87820**	**77983(-9)**

The Use of the Gegenbauer Polynomials

For the numerical evaluation of the finite-part singular integral (2.4.1) consider the following expression [41]:

$$G_n = \sum_{r=1}^{n} A_r g(x_r) \tag{2.4.53}$$

in which $x_1, x_2, \ldots, x_n$ are the zeros of the Gegenbauer polynomial $C_n^{\nu}(x)$ and also:

$$A_r = \int_{-1}^{1} \frac{C_n^{\nu}(x)w(x)\,\mathrm{d}x}{(x-x_r)\left[C_n^{\nu}(x_r)\right]'} \tag{2.4.54}$$

The integral $\int_{-1}^{1} g(x)\,\mathrm{d}x$ is estimated by (2.4.53), when $g(x)$ possesses no singularities in the interval $[-1,1]$. For the numerical evaluation of the finite-part singular integral (2.4.1) consider the following two cases:

First case. None of the poles z coincides with a zero x_r of $C_n^{\nu}(x)$. We consider the Cauchy-type integral:

$$\int_L \frac{g(z)w(z)\,\mathrm{d}z}{(z-x)C_n^{\nu}(x)} \tag{2.4.55}$$

in which L is a closed contour containing the interval $[-1,1]$.

Furthermore, Cauchy's residue theorem will be applied to the above integral, if of course, in addition to the poles $z_1, z_2, \ldots, z_t$, L contains further simple poles $z_t+1, \ldots, z_m$ of $g(z)$ with residues $\rho_{t+1}, \ldots, \rho_m$:

$$\begin{aligned} g(x) = \sum_{r=1}^{n} \frac{C_n^{\nu}(x)g(x_r)}{(x-x_r)\left[C_n^{\nu}(x_r)\right]'} + \sum_{r=1}^{m} \frac{\rho_r C_n^{\nu}(x)}{(x-z_r)C_n^{\nu}(z_r)} + \\ C_n^{\nu}(x)\int_L \frac{g(z)\,\mathrm{d}z}{(z-x)C_n^{\nu}(z)} \end{aligned} \tag{2.4.56}$$

Beyond the above, $g(x)$ is integrated along a path L with a small semi-circular indentation of radius ε to carry it above each of the poles z_r ($r = 1, 2, \ldots, t$). Letting $\varepsilon \to 0$, we obtain from (2.4.56):

$$\begin{aligned} \int_{-1}^{1} g(x)w(x)\,\mathrm{d}x = \int_{-1}^{1} \sum_{r=1}^{n} \frac{C_n^{\nu}(x)g(x_r)w(x)}{(x-x_r)\left[C_n^{\nu}(x_r)\right]'}\,\mathrm{d}x + \int_{-1}^{1} \sum_{r=1}^{m} \frac{\rho_r C_n^{\nu}(x)w(x)}{(x-z_r)C_n^{\nu}(z_r)}\,\mathrm{d}x \\ + \int_{-1}^{1} \left[C_n^{\nu}(x)\int_L \frac{g(z)w(x)\,\mathrm{d}z}{(z-x)C_n^{\nu}(z)} \right] \mathrm{d}x \end{aligned} \tag{2.4.57}$$

By using (2.4.54) one has:

$$\int_{-1}^{1} g(x)w(x)\,\mathrm{d}x = \sum_{r=1}^{n} A_r g(x_r) + \sum_{r=1}^{m} \frac{\rho_r}{C_n^v(z_r)} \int_{-1}^{1} \frac{C_n^v(x)w(x)}{x - z_r}\,\mathrm{d}x + \int_{-1}^{1} C_n^v(x)\,\mathrm{d}x \int_L \frac{g(z)w(x)}{(z-x)C_n^v(z)}\,\mathrm{d}z \tag{2.4.58}$$

Thus, by combining (2.4.53) and (2.4.58) we obtain:

$$\int_{-1}^{1} g(x)w(x)\,\mathrm{d}x = G_n + \sum_{r=1}^{m} \frac{\rho_r}{C_n^v(z_r)} \int_{-1}^{1} \frac{C_n^v(x)w(x)}{x - z_r}\,\mathrm{d}x + \int_{-1}^{1} C_n^v(x)\,\mathrm{d}x \int_L \frac{g(z)w(x)}{(z-x)C_n^v(z)}\,\mathrm{d}z \tag{2.4.59}$$

Also, if $z \in [-1,1]$, one has from [56, p.281, Eq. (5)]:

$$\int_{-1}^{1} (z-x)^{-1} x^m (1-x^2)^{v-1/2} C_n^v(x)\,\mathrm{d}x = M(z) P_{n+v-1/2}^{v-1/2}(z)$$

with: $m \le n$ and $\mathrm{Re}\,v > -1/2$ (2.4.60)

in which $P_{n+v-1/2}^{v-1/2}$ denotes the Gegenbauer polynomials of degree n+v-1/2 with weight function $w(x) = x^m (1-x^2)^{v-1/2}$ and:

$$M(z) = \frac{\pi^{1/2} 2^{3/2-v}}{\Gamma(v)} e^{-(v-1/2)\pi i} z^m (z^2 - 1)^{1/2v-1/4} \tag{2.4.61}$$

Moreover, by using (2.4.60) let us write (2.4.59) in the form:

$$\int_{-1}^{1} g(x)w(x)\,\mathrm{d}x = G_n - \sum_{r=1}^{m} \frac{\rho_r}{C_n^v(z_r)} M(z_r) P_{n+v-1/2}^{v-1/2}(z_r) + \int_L \frac{M(z) P_{n+v-1/2}^{v-1/2}(z) g(z)}{C_n^v(z)}\,\mathrm{d}z \tag{2.4.62}$$

Introducing the expressions:

$$\Lambda_n = G_n - R_n \tag{2.4.63}$$

$$R_n = \sum_{r=1}^{m} \rho_r M(z_r) \Omega_n(z_r) \tag{2.4.64}$$

where:

$$M(z_r) = \frac{\pi^{1/2} 2^{3/2-\nu}}{\Gamma(\nu)} e^{-(\nu-1/2)\pi i} z_r^m (z_r^2 - 1)^{1/2\nu - 1/4} \tag{2.4.65}$$

$$\Omega_n(z_r) = \frac{P_{n-\nu-1/2}^{\nu-1/2}(z_r)}{C_n^\nu(z_r)} \tag{2.4.66}$$

$$E_n = \int_L \frac{M(z) P_{n+\nu-1/2}^{\nu-1/2}(z) g(z)}{C_n^\nu(z)} \mathrm{d}z \tag{2.4.67}$$

we obtain:

$$\int_{-1}^{1} g(x) w(x) \mathrm{d}x = \Lambda_n + E_n$$

where E_n denotes the error of (2.4.62).

Second case. If some poles of $g(z)$ coincide with zeros of $C_n^\nu(z)$, then the equations must be modified. For this case a very specific method will be followed.

Consider a pole z_s, which coincides with a zero x_s of $C_n^\nu(z)$. Then it can be shown that the residue of $g(z)/\left[(z-x)C_n^\nu(x)\right]$ at x_s is:

$$\frac{1}{(x_s - x)\left[C_n^\nu(x)\right]'} \left[g_s - \frac{\rho_s x_s}{1 - x_s^2} - \frac{\rho_s}{x_s - x} \right] \tag{2.4.68}$$

where g_s is the constant term in the Laurent expression of $g(z)$ about x_s. Hence:

$$g_s = \lim_{x \to x_s} \frac{\mathrm{d}}{\mathrm{d}x}\left[(x - x_s) g(x)\right] \tag{2.4.69}$$

By inserting the residue in the argument used for the first case, then we find that the former expressions can still be used with the following modifications:

(a) The term G_n corresponding to $r = s$ in the summation in (2.4.53) must be replaced by $A_s g_s$ where A_s is given by (2.4.54).

(b) The term corresponding to $r = s$ in the summation of (2.4.64) must be replaced by $K_s \rho_s$ where:

$$K_s = \frac{A_s x_s}{2(1-x_s^2)} - \frac{1}{2\left[C_n^v(x_s)\right]'} \int_{-1}^{1} \frac{C_n^v(x) w(x) \, \mathrm{d}x}{(x_s - x)^2} \tag{2.4.70}$$

In order a simpler expression for K_s to be found, the Christoffel-Darboux formula is used [58, Eq.(22.12.1)]:

$$\sum_{m=0}^{n} \frac{1}{h_m} f_m(x) f_m(y) = \frac{K_n}{K_{n+1} h_n} \frac{f_{n+1}(x) f_n(y) - f_n(x) f_{n+1}(y)}{x - y} \tag{2.4.71}$$

For the this case the Christoffel-Darboux formula reduces to:

$$\frac{C_n^v(x)}{(x_s - x)^2} = \frac{1}{(n+1) C_{n+1}^v(x_s)} \sum_{k=0}^{n-1} \frac{(2k+1) C_k^v(x_s) C_k^v(x)}{x_s - x} \tag{2.4.72}$$

By integrating (2.4.72) one has:

$$\int_{-1}^{1} \frac{C_n^v(x) w(x)}{(x_s - x)^2} \mathrm{d}x = \frac{1}{(n+1) C_{n+1}^v(x_s)} \int_{-1}^{1} \sum_{k=0}^{n-1} \frac{(2k+1) C_k^v(x_s) C_k^v(x) w(x)}{x_s - x} \mathrm{d}x \tag{2.4.73}$$

Hence, (2.4.73) reduces to:

$$\int_{-1}^{1} \frac{C_n^v(x) w(x)}{(x_s - x)^2} \mathrm{d}x = \frac{1}{(n+1) C_{n+1}^v(x_s)} \sum_{k=0}^{n-1} (2k+1) C_k^v(x_s) \int_{-1}^{1} \frac{C_k^v(x) w(x)}{x_s - x} \mathrm{d}x \tag{2.4.74}$$

By combining (2.4.60) and (2.4.74) we get:

$$\int_{-1}^{1} \frac{C_n^v(x) w(x)}{(x_s - x)^2} \mathrm{d}x = \frac{1}{(n+1) C_{n+1}^v(x_s)} \sum_{k=0}^{n-1} (2k+1) C_k^v(x_s) M(x_s) P_{n+v-1/2}^{v-1/2}(x_s) \tag{2.4.75}$$

Beyond the above, from (2.4.70) and (2.4.72) one obtains:

$$K_s = \frac{A_s x_s}{2(1-x_s^2)} - \frac{1}{2\left[C_n^v(x_s)\right]' (n+1) C_{n+1}^v(x_s)} \sum_{k=0}^{n-1} (2k+1) C_n^v(x_s) M(x_s) P_{n+v-1/2}^{v-1/2}(x_s) \tag{2.4.76}$$

Some additional notes for the use of the Gegenbauer polynomials:

First Note

By replacing (2.4.60) with eq. (5) in [57, p.281] we obtain for the first case of the previous Section:

$$\int_{-1}^{1}(z-x)^{-1}x^{n+1}(1-x^2)^{v-1/2}C_n^v(x)\,\mathrm{d}x=$$

$$\frac{\pi^{1/2}2^{3/2-v}}{\Gamma(v)}e^{-(v-1/2)\pi i}z^{n+1}(z^2-1)^{1/2v-1/4}P_{n+v-1/2}^{v-1/2}(z)$$
$$-\frac{\pi 2^{1-2v-n}n!}{\Gamma(v)\Gamma(v+n+1)}\quad(\mathrm{Re}\,v>1/2)\tag{2.4.77}$$

with the weight function:

$$w(x)=x^{n+1}(1-x^2)^{v-1/2}\tag{2.4.78}$$

Abbreviating:

$$N(z)=\frac{\pi^{1/2}2^{3/2-v}}{\Gamma(v)}e^{-(v-1/2)\pi i}z^{n+1}(z^2-1)^{1/2v-1/4}\tag{2.4.79}$$

and:

$$A_n=\frac{\pi 2^{1-2v-n}n!}{\Gamma(v)\Gamma(v+n+1)}\tag{2.4.80}$$

one has from (2.4.77):

$$\int_{-1}^{1}(z-x)^{-1}w(x)C_n^v(x)\,\mathrm{d}x=N(z)P_{n+v-1/2}^{v-1/2}(z)-A_n\tag{2.4.81}$$

Hence, by using (2.4.81) then (2.4.59) can be written in the form:

$$\int_{-1}^{1}g(x)w(x)\,\mathrm{d}x=G_n-\sum_{r=1}^{m}\frac{\rho_r}{C_n^v(z_r)}\left[N(z_r)P_{n+v-1/2}^{v-1/2}(z_r)-A_n\right]$$
$$+\int_L\frac{\left[N(z)P_{n+v-1/2}^{v-1/2}(z)-A_n\right]g(z)}{C_n^v(z)}\mathrm{d}z\tag{2.4.82}$$

If we put:

$$\Lambda_n=G_n-R_n\tag{2.4.83}$$

$$R_n = \sum_{r=1}^{m} \frac{\rho_r}{C_n^v(z_r)} \left[N(z_r) P_{n+v-1/2}^{v-1/2}(z_r) - A_n \right] \tag{2.4.84}$$

$$N(z_r) = \frac{\pi^{1/2} 2^{3/2-v}}{\Gamma(v)} e^{-(v-1/2)\pi i} z_r^{n+1} (z_r^2 - 1)^{1/2v-1/4} \tag{2.4.85}$$

and abbreviate the error function by:

$$E_n = \int_L \frac{\left[N(z) P_{n+v-1/2}^{v-1/2}(z) - A_n \right] g(z)}{C_n^v(z)} \mathrm{d}z \tag{2.4.86}$$

we get:

$$\int_{-1}^{1} g(x) w(x) \mathrm{d}x = \Lambda_n + E_n$$

Second Note

If we now replace (2.4.60) by eq. (17) in [57, p.283], one has:

$$\int_{-1}^{1} (z-x)^{-1} (1-x^2)^{v-1/2} C_m^v(x) C_n^v(x) \mathrm{d}x$$

$$= \frac{\pi^{1/2} 2^{3/2-v}}{\Gamma(v)} e^{-(v-1/2)\pi i} (z^2-1)^{1/2v-1/4} C_m^v P_{n+v-1/2}^{v-1/2}(z) \tag{2.4.87}$$

$(m \le n, \mathrm{Re}\, v > -1/2)$

with weight function:

$$w(x) = (1-x^2)^{v-1/2} \tag{2.4.88}$$

With:

$$N_1(z) = \frac{\pi^{1/2} 2^{3/2-v}}{\Gamma(v)} e^{-(v-1/2)\pi i} (z^2-1)^{1/2v-1/4} \tag{2.4.89}$$

eq. (2.4.87) takes the form:

$$\int_{-1}^{1} (z-x)^{-1} w(x) C_m^v(x) C_n^v(x) \mathrm{d}x = N_1(z) C_m^v(z) P_{n+v-1/2}^{v-1/2}(z) \tag{2.4.90}$$

Accordingly, by using (2.4.60) we write (2.4.59) in the form:

$$\int_{-1}^{1} g(x)w(x)\,\mathrm{d}x = G_n - \sum_{r=1}^{m} \frac{\rho_r}{C_n^v(z_r)} \left[N_1(z_r) C_m^v(z_r) P_{n+v-1/2}^{v-1/2}(z_r) \right]$$

$$+ \int_{L} \frac{N_1(z) C_m^v(z) P_{n+v-1/2}^{v-1/2} g(z)}{C_n^v(z)} \mathrm{d}z \tag{2.4.91}$$

The General Type of the Finite-Part Singular Integrals

Furthermore, the numerical behaviour of the finite part singular integrals is investigated. From (2.4.62) we obtain the numerical evaluation for the Cauchy-type singular integral:

$$\Phi(z) = \int_{-1}^{1} \frac{g(t)w(t)}{t-z} \mathrm{d}t = \sum_{r=1}^{m} A_r \frac{g(t_r)}{t_r - z} - \sum_{r=1}^{m} \frac{\rho_r M(z_r)}{t - z_r} \frac{P_{n+v-1/2}^{v-1/2}(z_r)}{C_n^v(z_r)} + E_n \tag{2.4.92}$$

For the generalized type of the finite part integral, by differentiating (2.4.92) with respect to z, one has:

$$⨎_{-1}^{1} \frac{g(t)w(t)}{(t-z)^2} \mathrm{d}t \cong \sum_{r=1}^{m} A_r \frac{g(t_r)}{(t_r - z)^2} - \frac{\mathrm{d}}{\mathrm{d}z} \left[\sum_{r=1}^{m} \frac{\rho_r M(z_r)}{t - z_r} \frac{P_{n+v-1/2}^{v-1/2}(z_r)}{C_n^v(z_r)} \right] \tag{2.4.93}$$

By a second differentiation of (2.7.93) with respect to z follows:

$$⨎_{-1}^{1} \frac{2g(t)w(t)}{(t-z)^3} \mathrm{d}t \cong \sum_{r=1}^{m} A_r \frac{2g(t_r)}{(t_r - z)^3} - \frac{\mathrm{d}^2}{\mathrm{d}z^2} \left[\sum_{r=1}^{m} \frac{\rho_r M(z_r)}{t - z_r} \frac{P_{n+v-1/2}^{v-1/2}(z_r)}{C_n^v(z_r)} \right] \tag{2.4.94}$$

Differentiating (μ-1) times one obtains:

$$(\mu-1)!⨎_{-1}^{1} \frac{g(t)w(t)}{(t-z)^\mu} \mathrm{d}t \cong \sum_{r=1}^{m} A_r \frac{(\mu-1)!\,g(t_r)}{(t_r - z)^\mu} - \frac{\mathrm{d}^{(\mu-1)}}{\mathrm{d}z^{(\mu-1)}} \left[\sum_{r=1}^{m} \frac{\rho_r M(z_r)}{t - z_r} \frac{P_{n+v-1/2}^{v-1/2}(z_r)}{C_n^v(z_r)} \right] \tag{2.4.95}$$

On the other hand, it is known that $\Gamma(\mu) = (\mu\text{-}1)!$ for $\mu = 1, 2, 3, \ldots$ So (2.4.95) takes the final form:

$$\Phi(z,\mu) = \Gamma(\mu) \oint_{-1}^{1} \frac{g(t)w(t)}{(t-z)^{\mu}} \mathrm{d}t \cong \Gamma(\mu) \sum_{r=1}^{m} A_r \frac{g(t_r)}{(t_r - z)^{\mu}} - \frac{\mathrm{d}^{(\mu-1)}}{\mathrm{d}z^{(\mu-1)}} \left[\sum_{r=1}^{m} \frac{\rho_r M(z_r)}{t - z_r} \frac{P_{n+v-1/2}^{v-1/2}(z_r)}{C_n^{v}(z_r)} \right] \tag{2.4.96}$$

Hence, (2.4.96) gives the new integration rule for the finite-part singular integral. Another interesting method for the numerical evaluation of the finite-part integrals was proposed by H. R. Kutt [4].

He has expanded the function $w(x)\varphi(x)$ of the finite-part integral (2.4.1) in its Taylor series. The main disadvantage of his method lies in the fact, that the error function in his formula seems to be very complicated.

2.5. Numerical Integration Rules for Finite-Part Singular Integrals and Integral Equations with Logarithmic Singularities

Let the finite-part singular integral: [11]

$$\Phi(z,\mu) = \Gamma(\mu) \oint_{0}^{1} \frac{\ln(1/t)\varphi(t)}{(t-z)^{\mu}} \mathrm{d}t, \quad \mu = 1,2,3... \tag{2.5.1}$$

in which $\ln(1/t)$ denotes the logarithmic weight function, defined for every $t \in [0,1]$, $\varphi(t)$ an analytic function of t in any plane domain S, containing the interval $[0,1]$ and $\Gamma(\mu)$ the Gamma function.

Furthermore, the finite-part singular integral equation of the second kind with logarithmic singularities is defined as:

$$A(z)\ln(1/z)g(z) + \Gamma(\mu) \oint_{0}^{1} B(t) \frac{\ln(1/t)g(t)}{(t-z)^{\mu}} \mathrm{d}t + \int_{0}^{1} K(t,z)\ln(1/t)g(t)\,\mathrm{d}t = f(z,\mu) \tag{2.5.2}$$

$(\mu = 1,2,3...)$

where $A(z)$, $B(t)$ and $f(z,\mu)$ are known functions Hölder-continuous in the closed interval $[0,1]$, $K(t,z)$ is the Fredholm kernel, $\Gamma(\mu)$ the Gamma function and $g(z)$ the unknown function. When $A(z)=0$ and $B(t)=1$, then (2.5.2) is said to be the finite-part singular integral equation of the first kind.

Beyond the above, the finite-part singular integral (2.5.1) will be calculated numerically for the special case $\mu = 1$:

$$\Phi(z)=\int_L \frac{\varphi(t)w(t)}{(t-z)}\mathrm{d}t \cong \sum_{k=1}^{m} A_k \frac{\varphi(t_k)}{t_k - z} - 2\varphi(z)\frac{\lambda_n(z)}{J_n(z)} \tag{2.5.3}$$

in which the functions $\lambda_n(z)$, $J_n(z)$, are defined as:

$$J_n(z)=\prod_{k=1}^{n}(z-t_k) \tag{2.5.4}$$

$$\lambda_n(z)=-\frac{1}{2}\int_L \frac{w(t)J_n(t)}{t-z}\mathrm{d}t \tag{2.5.5}$$

where t_k denotes the abscissae and A_k the weights of the given integration rule.

By differentiating eq. (2.5.3) with respect to z, $(\mu - 1)$ times one obtains:

$$(\mu-1)! \oint \frac{\varphi(t)w(t)}{(t-z)^{\mu}}\mathrm{d}t \cong \sum_{k=1}^{n} A_k \frac{(\mu-1)!\varphi(t_k)}{(t_k - z)^{\mu}} - 2\frac{\mathrm{d}^{(\mu-1)}}{\mathrm{d}z^{(\mu-1)}}\left(\varphi(z)\frac{\lambda_n(z)}{J_n(z)}\right) \tag{2.5.6}$$

or:

$$\oint_L \frac{\varphi(t)w(t)}{(t-z)^{\mu}}\mathrm{d}t \cong \sum_{k=1}^{n} A_k \frac{\varphi(t_k)}{(t_k - z)^{\mu}} - \frac{2}{(\mu-1)!}\frac{\mathrm{d}^{(\mu-1)}}{\mathrm{d}z^{(\mu-1)}}\left(\varphi(z)\frac{\lambda_n(z)}{J_n(z)}\right) \tag{2.5.7}$$

and finally one has:

$$\Phi(z,\mu)=\Gamma(\mu)\int_L \frac{\varphi(t)w(t)}{(t-z)^{\mu}}\mathrm{d}t \cong$$
$$\Gamma(\mu)\sum_{k=1}^{n} A_k \frac{\varphi(t_k)}{(t_k - z)^{\mu}} - \frac{\mathrm{d}^{(\mu-1)}}{\mathrm{d}z^{(\mu-1)}}\left(\varphi(z)\frac{\lambda_n(z)}{J_n(z)}\right) \tag{2.5.8}$$

Hence, the complete approximation formula for the finite-part singular integral (2.8.1) with weight function $w(t)$, is defined by eq. (2.5.8).

Computational Integration Rules with Logarithmic Singularities

For the numerical evaluation of (2.5.1) some orthogonal polynomials are used, defined by the following recurrence formula [11], [56]:

$$P_0(z)=1, \qquad P_1(z)=z-b_1$$

$$P_n(z)=(z-b_n)P_{n-1}(z)-c_n P_{n-2}(z), \quad n \geq 2 \tag{2.5.9}$$

As the coefficients b_n, c_n and the polynomials themselves for $n > 4$, are not known exactly, are extremely difficult to be computed. In [56] A. H. Stroud and D. Secrest gave the exact values for $n = 1, 2, 3, 4$ and approximate values to 30 significant figures for $n = 1$ to 16.

The approximate coefficients in the polynomials are listed with the constant term first:

$$b_2 = \frac{13}{28} \quad c_2 = \frac{7}{144}$$

$$b_3 = \frac{8795}{18116} \quad c_3 = \frac{647}{11025}$$

$$b_4 = \frac{124351943}{252694908} \quad c_4 = \frac{71180289}{1172105200}$$

$$P_0 = 1$$

$$P_1 = z - \frac{1}{4}$$

$$P_2 = z^2 - \frac{5}{7}z + \frac{17}{252}$$

$$P_3 = z^3 - \frac{3105}{2588}z^2 + \frac{178281}{501425}z - \frac{4679}{258800}$$

$$P_4 = z^4 - \frac{165196}{97641}z^3 + \frac{67227}{75943}z^2 - \frac{79564}{531601}z + \frac{2296639}{478440900}$$

Thus, eq. (2.5.5) takes the following form:

$$\lambda_n(z) = -\frac{1}{2}\int_0^1 \ln\left(\frac{1}{t}\right)\frac{P_n(t)}{t-z}\,\mathrm{d}t \tag{2.5.10}$$

and because of (2.5.9) we obtain:

$$\lambda_n(z) = (z - b_n)\lambda_{n-1}(z) - c_n\lambda_{n-2}(z), \quad n \geq 2 \tag{2.5.11}$$

Hence, for the cases $n = 1, 2, 3, 4$, relation (2.5.11) reduces to:

$$\lambda_1(z) = (z - b_n)\lambda_0(z) - \frac{1}{2} \tag{2.5.12}$$

$$\lambda_2(z) = (z - b_n)\lambda_1(z) - c_n\lambda_0(z) \tag{2.5.13}$$

$$\lambda_3(z) = (z - b_n)\lambda_2(z) - c_n\lambda_1(z) \tag{2.5.14}$$

$$\lambda_4(z) = (z - b_n)\lambda_3(z) - c_n\lambda_2(z) \tag{2.5.15}$$

Beyond the above, according to the definition of the principle value of the integral, one has for the function $\lambda_0(z)$ defined on the interval [0,1]:

$$\lambda_0(z)=\frac{1}{2}\lim_{\varepsilon\to 0}\left(\int_0^{x-\varepsilon}\frac{\ln t}{t-x}\,\mathrm{d}t+\int_{x+\varepsilon}^{1}\frac{\ln t}{t-x}\,\mathrm{d}t\right),\quad 0<x<1 \tag{2.5.16}$$

Thus, we have:

$$2\lambda_0(x)=\ln|x|\ln\left|\frac{1-x}{x}\right|-D\left(\frac{x-1}{x}\right)+D(1)-\lim_{\varepsilon\to 0}\left(D\left(\frac{\varepsilon}{x}\right)-D\left(-\frac{\varepsilon}{x}\right)\right), \tag{2.5.17}$$

$$(0<x<1)$$

in which:

$$D(z)=-\int_0^z\frac{\ln(1-u)}{u}\,\mathrm{d}u \tag{2.5.18}$$

denotes the dilogarithm function.

Moreover, the limit $\lim_{\varepsilon\to 0}\left(D\left(\frac{\varepsilon}{x}\right)-D\left(-\frac{\varepsilon}{x}\right)\right)$ is equal to zero, since the power-series expansion of the dilogarithm function $D(z)=\sum_{n=1}^{\infty}\left(\frac{z^n}{n^2}\right)$ converges absolutely and uniformly for $|z|<1$:

$$2\lambda_0(x)=\ln|x|\ln\left|\frac{1-x}{x}\right|-D\left(\frac{x-1}{x}\right)+D(1) \tag{2.5.19}$$

which holds also for the case where $x<0$ or $x>1$. Therefore, by replacing $D(1)$ and $D\left[\frac{x-1}{x}\right]$ by their expressions given by [59, eq (2.3) and (5.4)] one obtains:

$$2\lambda_0(x)=\begin{cases}-D(x)-\frac{1}{2}\ln^2|x|+\frac{\pi^2}{3}, & for\, x>0\\ -D(x)-\frac{1}{2}\ln^2|x|-\frac{\pi^2}{6}, & for\, x<0\end{cases} \tag{2.5.20}$$

Also, if z is complex, then eq. (2.5.19) reduces to:

$$2\lambda_0(z)=-D(z)-\frac{1}{2}\ln^2(-z)-\frac{\pi^2}{6} \tag{2.5.21}$$

Equations (2.5.19), (2.5.20) and (2.5.21), define completely the functions $\lambda_n(z)$ in the whole plane. In general, quadrature rule of the open (Gauss), semiopen (Radau) and closed (Lobatto) types are used [11], for the numerical evaluation of the singular integral equations. On the other hand, the polynomials $P_n(z)$ and the corresponding functions $\lambda_n(z)$ are used only for the Gauss method.

In the Radau method the appropriate polynomials are given by:

$$J_n(z) = P_n(z) + d_n P_{n-1}(z), \quad n \geq 1 \tag{2.5.22}$$

in which:

$$d_n = -\frac{P_n(\rho)}{P_{n-1}(\rho)}, \quad \rho = 0 \text{ or } 1 \tag{2.5.23}$$

Also, in the Lobatto method the appropriate polynomials are equal to:

$$J_n(z) = P_n(z) + d_n P_{n-1}(z) + e_n P_{n-2}(z), \quad n \geq 2 \tag{2.5.24}$$

where:

$$\begin{aligned} d_n &= -(P_n(0)P_{n-2}(1) - P_n(1)P_{n-2}(0))/D \\ e_n &= -(P_n(0)P_{n-1}(1) - P_n(1)P_{n-1}(0))/D \\ D &= P_{n-1}(0)P_{n-2}(1) - P_{n-1}(1)P_{n-2}(0) \end{aligned}$$

The weights of the n-point quadrature rules for these three methods are:

$$A_k = -2\frac{\lambda_n(t_k)}{J_n'(t_k)}, \quad k = 1,2,\ldots,n \tag{2.5.25}$$

in which $\lambda_n(z)$ and $J_n(z)$ denote the functions corresponding to the method employed and t_k are the roots of $J_n(z)$. Hence, the singular integral (2.5.1) for $\mu = 1$ can be evaluated by using the following quadrature formula [11]:

$$I(z) = \int_0^1 \ln\left(\frac{1}{t}\right)\frac{\varphi(t)}{t-z}\,\mathrm{d}t = \sum_{k=1}^{n} A_k \frac{\varphi(t)}{t_k - z} - 2\sum_{k=1}^{m} \frac{e_k \lambda_n(z_k)}{J_n(z_k)} + E_n \tag{2.5.26}$$

where the error E_n is given by:

$$E_n = \frac{1}{\pi i}\int_L \frac{\lambda_n(r)}{J_n(r)}\frac{\varphi(r)}{r-z}\,\mathrm{d}r \tag{2.5.27}$$

By differentiation of (2.5.26) (μ-1) times, we obtain:

$$(\mu-1)!\underset{0}{\overset{1}{\oint}}\ln\left(\frac{1}{t}\right)\frac{\varphi(t)}{(t-z)^{\mu}}\mathrm{d}t \cong \sum_{k=1}^{n} A_k \frac{(\mu-1)!\varphi(t_k)}{(t_k-z)^{\mu}} - 2\sum_{k=1}^{m} e_k \frac{\mathrm{d}^{(\mu-1)}}{\mathrm{d}z^{(\mu-1)}}\left(\frac{\lambda_n(z_k)}{J_n(z_k)}\right) \tag{2.5.28}$$

Finally, the finite-part singular integral with logarithmic singularities is numerically approximated by the following formula:

$$\begin{aligned}\Phi(z,\mu) = \Gamma(\mu)\underset{0}{\overset{1}{\oint}}\ln\left(\frac{1}{t}\right)\frac{\varphi(t)}{(t-z)^{\mu}}\mathrm{d}t &\cong \Gamma(\mu)\sum_{k=1}^{n} A_k \frac{\varphi(t_k)}{(t_k-z)^{\mu}} \\ &- 2\sum_{k=1}^{m} e_k \frac{\mathrm{d}^{(\mu-1)}}{\mathrm{d}z^{(\mu-1)}}\left(\frac{\lambda_n(z_k)}{J_n(z_k)}\right)\end{aligned} \tag{2.5.29}$$

In the above expression z_k are the simple poles of $\Phi(z,\mu)$, lying in the interior of a simple-closed curve C surrounding the integration interval [0,1].

2.6. Approximation Methods for Singular Integral Equations on Lyapounov Contours in Banach Spaces

Several approximation methods are further investigated for the numerical evaluation of the singular integral equations on Lyapounov contours, by using the Faber polynomials. Such singular integral equations are defined in Banach spaces and the Faber-Laurent expansion is used for their numerical evaluation [18].

Thus, a general approximation method is examined together with its connection with algorithms and general schemes are introduced to the approximate solution of singular integral equations on Lyapounov contours. Particular attention has been concentrated on the construction and verification of algorithms which are effective when the integration interval is not a smooth contour like in the previous sections, but a Lyapounov contour.

By the following theorems is proved the approximation of functions in a complex domain, when these functions are defined in Banach spaces $H_\gamma(\Gamma)$, of functions satisfying Hölder's condition with exponent γ $(0<\gamma\leq 1)$ in which Γ denotes a closed Lyapounov contour.

Definition 2.6.1.

Consider a function $h(t)$ defined in the Banach space $H_\gamma(\Gamma)$, $t\in\Gamma$, of functions satisfying Hölder's condition with exponent γ $(0<\gamma\leq 1)$ [25], where Γ denotes a closed Lyapounov contour which bounds a simply connected domain containing the point $t=0$. Then, for the function $h(t)\in H_\gamma(\Gamma)$ the following Faber-Laurent expansion exists: [60]

$$h(t)=\sum_{k=o}^{\infty} e_k \Phi_k(t)+\sum_{k=1}^{\infty} l_k F_k(1/t) \tag{2.6.1}$$

in which $\Phi_k(t)$ and $F_k(1/t)$ are Faber polynomials and e_k and l_k the Faber-Laurent coefficients of $h(t)$.

Definition 2.6.2.

By E_n is defined the operator mapping each $h(t)\in H_\gamma(\Gamma)$ into the n-th partial sum of the Faber-Laurent series:[60]

$$(E_n h)(t)=\sum_{k=o}^{n} e_k \Phi_k(t)+\sum_{k=1}^{n} l_k F_k(1/t), \quad t\in\Gamma \tag{2.6.2}$$

where $\Phi_k(t)$ and $F_k(1/t)$ are the Faber polynomials and e_k and l_k the Faber-Laurent coefficients of $h(t)$.

Lemma 2.6.1.

Let by P_n the projection operator on $H_\gamma(\Gamma)$ equal to:

$$(P_n h)(t)=\sum_{j=-n}^{n} h_j t^j \tag{2.6.3}$$

in which $h(t)\in H_\gamma(\Gamma)$, $t\in\Gamma$ and the complex sequence $\{h_j\}$, j=0,±1, ±2,±.... is defined as:

$$h_j=\frac{1}{j!}\frac{\mathrm{d}^j[Ah(t)]}{\mathrm{d}t^j}\bigg|_{t=0}, \quad (j=0,1,2,.....)$$

$$h_j=\frac{1}{(-j)!}\frac{\mathrm{d}^{(-j)}[Bh(t^{-1})]}{\mathrm{d}t^{(-j)}}\bigg|_{t=0}, \quad (j=-1,-2,.....) \tag{2.6.4}$$

where $A=\dfrac{(I+E)}{2}$ and $B=\dfrac{(I-E)}{2}$, with I the identity operator and E the Cauchy singular integral:

$$(Eu)(t)=\frac{1}{\pi i}\int_\Gamma \frac{u(x)}{x-t}\,\mathrm{d}x, \quad t\in\Gamma \tag{2.6.5}$$

Then, the following formula is valid:

$$h(t)-(P_n h)(t)=h(t)-(E_n h)(t)-\frac{1}{\pi i}\int_\Gamma [(Ah)(x)-(E_n Ah)(x)\cdot \\ \cdot\sum_{j=0}^{n}\frac{t^j}{x^{j+1}}\,\mathrm{d}x-\frac{1}{\pi i}\int_\Gamma [(Bh)(x)-(E_n Bh)(x)]\sum_{j=0}^{n-1}\frac{x^j}{t^{j+1}}\,\mathrm{d}x, \quad t\in\Gamma \tag{2.6.6}$$

Proof. Consider the Faber polynomials for the Lyapounov contour Γ:

$$\Phi_k(t) = \sum_{j=0}^{k} \delta_j t^j$$

$$F_k\left(\frac{1}{t}\right) = \sum_{j=1}^{k} \theta_j t^{-j} \tag{2.6.7}$$

Thus, we have:

$$(P_n \Phi_k)(t) = \begin{Bmatrix} \Phi_k(t), \ k \le n \\ \sum_{j=0}^{n} \delta_j t^j, \ k > n \end{Bmatrix} \tag{2.6.8}$$

and:

$$(P_n F_k)(t) = \begin{Bmatrix} F_k\left(1/t\right), \ k \le n \\ \sum_{j=1}^{n} \theta_j t^{-j}, \ k > n \end{Bmatrix} \tag{2.6.9}$$

Therefore, by using (2.6.1) we obtain:

$$(P_n h)(t) = P_n\left(\sum_{k=0}^{\infty} e_k \Phi_k(t)\right) + P_n\left(\sum_{k=1}^{\infty} l_k F_k\left(\frac{1}{t}\right)\right) \tag{2.6.10}$$

Furthermore, by using (2.6.8) and (2.6.9), then (2.6.10) reduces to:

$$(P_n h)(t) = (E_n h)(t) + \sum_{k=n+1}^{\infty} e_k \sum_{j=0}^{n} \frac{1}{\pi i} \int_\Gamma \frac{\Phi_k(x) t^j}{x^{j+1}} \mathrm{d}x +$$

$$\sum_{k=n+1}^{\infty} l_k \sum_{j=1}^{n} \frac{1}{\pi i} \int_\Gamma \frac{F_k(1/x) x^{j+1}}{t^j} \mathrm{d}x \tag{2.6.11}$$

Finally (2.6.1) can be written as:

$$(P_n h)(t) = (E_n h)(t) + \sum_{j=0}^{n} \frac{1}{\pi i} \int_\Gamma \sum_{k=n+1}^{\infty} e_k \frac{\Phi_k(x) t^j}{x^{j+1}} \mathrm{d}x +$$

$$+ \sum_{j=1}^{n} \frac{1}{\pi i} \int_\Gamma \sum_{k=n+1}^{\infty} l_k \frac{F_k(1/x) x^{j+1}}{t^j} \mathrm{d}x \tag{2.6.12}$$

Beyond the above, the following relations are obvious:

$$\sum_{k=n+1}^{\infty} e_k \Phi_k(t) = (Ah)(t) - (E_n Ah)(t) \tag{2.6.13}$$

and:

$$\sum_{k=n+1}^{\infty} l_k F_k(1/t) = (Bh)(t) - (E_n Bh)(t) \tag{2.6.14}$$

Hence, from (2.6.12), (2.6.13) and (2.6.14) follows the required (2.6.6) and thus, the proof of Lemma 2.6.1 is completed.

Theorem 2.6.1

Consider the function $h(t) \in H_\gamma(\Gamma)$, $t \in \Gamma$, P_n the projection operator on $H_\gamma(\Gamma)$ be given by (2.6.3), $R_n(h;\Gamma)$ the best uniform approximation of $h(t)$ on Γ by the following polynomials:

$$u_n(t) = \sum_{k=-n}^{n} \lambda_k t^k, \quad t \in \Gamma \tag{2.6.15}$$

and $A = \dfrac{(I+E)}{2}$, $B = \dfrac{(I-E)}{2}$, with I the identity operator and E the Cauchy singular integral (2.6.5).

Then, the following inequality is valid:

$$\begin{aligned} &|h(t) - (P_n h)(t)| \le c_1 \ln n R_n(h;\Gamma) + \\ &+ c_2 \ln^2 n R_n(Ah;\Gamma) + c_3 \ln^2 n R_n(Bh;\Gamma) \end{aligned} \tag{2.6.16}$$

in which c_1, c_2 and c_3 are constants not depending on n.

Proof. Since $h(t) \in H_\gamma(\Gamma)$, then one has:[60]

$$|h(t) - (E_n h)(t)| \le c_4 \ln n R_n(h;\Gamma) \tag{2.6.17}$$

where c_4 denotes a constant not depending on n.

Furthermore, $(Ah)(t)$ and $(Bh)(t)$ are in $H_\gamma(\Gamma)$ and thus by combining (2.6.6) and (2.6.17) we obtain the following inequality:

$$|h(t) - (P_n h)(t)| \le c_4 \ln n \left\{ R_n(h;\Gamma) + R_n(Ah;\Gamma) \frac{1}{\pi} \int_\Gamma \left| \sum_{j=0}^{n} \frac{t^j}{x^{j+1}} \right| |\mathrm{d}x| + \right.$$

$$+R_n(Bh;\Gamma)\frac{1}{\pi}\int_\Gamma\left|\sum_{j=1}^{n-1}\frac{x^j}{t^{j+1}}\right||\mathrm{d}\,x|\Bigg\} \tag{2.6.18}$$

Also, from [60] follows that:

$$\int_\Gamma\left|\sum_{j=0}^{n}\frac{t^j}{x^{j+1}}\right||\mathrm{d}\,x|\le c_5\ln n \tag{2.6.19}$$

$$\int_\Gamma\left|\sum_{j=0}^{n-1}\frac{x^j}{t^{j+1}}\right||\mathrm{d}\,x|\le c_6\ln n$$

Therefore, finally from (2.6.18) and (2.6.19) follows the required inequality (2.6.16).

Theorem 2.6.2.

Let the function $h(t)\in H_\varepsilon(\Gamma)$, $0\le\gamma\le\varepsilon\le 1$, *Pn* the projection operator on $H_\gamma(\Gamma)$ be given by (2.6.3) and c_7 a constant not depending on n. Then, the following inequality is valid:

$$\|h-P_nh\|_\gamma\le c_7\ln^2 n\cdot n^{\gamma-\varepsilon}H(h;\varepsilon) \tag{2.6.20}$$

Proof. In order to prove inequality (2.6.20), let us find a bound for $\max_{t\in\Gamma}|h(t)\text{-}(P_nh)(t)|$. Thus, from (2.6.16) follows:

$$\max_{t\in\Gamma}|h(t)-(P_nh)(t)|\le c_4c_8\ln n\cdot n^{-\varepsilon}[c_1H(h;\varepsilon)+$$

$$+c_2\ln nH(\mathrm{A}h;\varepsilon)+c_3\ln nH(Bh;\varepsilon)]\le c_9\ln^2 n\cdot n^{-\varepsilon}H(h;\varepsilon) \tag{2.6.21}$$

Hence, the required inequality (2.6.20) follows immediately from (2.6.21).

Lemma 2.6.2

Consider the function $h(t)\in H_\gamma(\Gamma)$, P_n the projection operator on $H_\gamma(\Gamma)$ be given by (2.6.3) and c_7 a constant not depending on n. Then, the following inequality exists:

$$\|P_nh\|_\gamma\le(1+c_7\ln^2 n)\|h\|_\gamma \tag{2.6.22}$$

Proof. Inequality (2.6.22) follows from (2.6.20), by replacing $\gamma=\varepsilon$.

Furthermore, these results are used for the proof of the existence and uniqueness of solutions for the linear equation systems, on which the singular integral equations are reduced. Also, the rate of convergence of the proposed method is also given, by following a special proof.

Existence theorems for singular integral equations

Definition 2.6.3
Let the singular integral equation:

$$\Gamma u \equiv a(t)u(t) + \frac{b(t)}{\pi i}\int_{\Gamma}\frac{u(x)}{x-t}dx + \int_{\Gamma}K(t,x)u(x)dx = f(t),\ \ t \in \Gamma \tag{2.6.23}$$

defined over the closed Lyapounov contour Γ, which bounds a simply connected domain containing the point $t=0$, $a(t)$, $b(t)$, $f(t)$ and $K(t,x)$ are known functions on the Banach space $H_{\gamma}(\Gamma)$ and $u(t)$ denotes the unknown function.

Theorem 2.6.3
Consider the singular integral equation (2.6.23) defined over the closed Lyapounov contour Γ. An approximate solution of this equation is of the form:

$$u_n(t) = \sum_{k=-n}^{n} \varepsilon_k t^k,\ \ t \in \Gamma \tag{2.6.24}$$

in which the unknown ε_k ($k = \overline{-n, n}$) satisfy the following linear equation system:

$$\sum_{k=0}^{n} M_{j-k}\varepsilon_k + \sum_{k=-n}^{-1} \Lambda_{j-k}\varepsilon_k + \sum_{k=-n}^{n} N_{jk}e_k = f_j \tag{2.6.25}$$

where M_j, Λ_j, N_{jk} and f_j, ($j=0,\pm1, \pm2, ...$, $k = \overline{-n, n}$) are calculated from (2.6.4), by replacing $h(t)$ by $M(t)=a(t)+b(t)$, $\Lambda(t) = a(t) - b(t), K(t,x) = \int_{\Gamma} k(t,x)x^k\,\mathrm{d}x$ and $f(t)$, respectively.

Then, if $M(t)\ \Lambda(t) \neq 0$ and $M(t)$, $\Lambda(t)$ $f(t)$ and $k(t,x)$ are in the space $H_{\varepsilon}(\Gamma)(0<\gamma<\varepsilon\leq 1)$, the linear system (2.6.25) has a unique solution ε_k. Also, the approximate solution (2.6.24) converges for $n\to\infty$, to the exact solution $u(t)$ of (2.6.23):

$$\lim_{n\to\infty} u_n(t) = u(t) \tag{2.6.26}$$

and the rate of convergence is given by the inequality:

$$\left\| u - u_n \right\|_{\gamma} \leq c \frac{\ln^2 n}{n^{\xi(\varepsilon)-\gamma}} H(u;\xi(\varepsilon)) \tag{2.6.27}$$

with:

$$\xi(\varepsilon) = \begin{cases} \varepsilon,\ \varepsilon < 1 \\ 1-\delta,\ \varepsilon = 1 \end{cases} \tag{2.6.28}$$

in which δ denotes an arbitrarily small positive constant and c a constant not depending on n.

Proof. The linear equation system (2.6.25) can be written by the following operator-form:

$$\Pi_n u_n \equiv P_n(MA + \Lambda B + K)P_n u_n(t) = (P_n f)(t), \quad t \in \Gamma \tag{2.6.29}$$

where $A = \frac{(I+E)}{2}$ and $B = \frac{(I-E)}{2}$, with I the identity operator, E the Cauchy singular integral (2.6.5) and P_n the projection operator on $H_\gamma(\Gamma)$, which is given by (2.6.3).

Beyond the above, it will be proved that for sufficiently large n, the operator Π_n has an inverse in $H_\gamma(\Gamma)$. Hence consider by $M(t) = M_-(t)M_+(t)$ and $\Lambda(t) = \Lambda_-(t)\,\Lambda_+(t)$ the canonical factorizations of $M(t)$ and $\Lambda(t)$.

Thus, one has: [61]

$$MA + \Lambda B + K = R_1 + R_2 \tag{2.6.30}$$

in which:

$$R_1 = (M_+ A + \Lambda_- B)(AM_- + B\Lambda_+) \tag{2.6.31}$$

and:

$$R_2 = M_+ BM_- A + \Lambda_- A\Lambda_+ B + K \tag{2.6.32}$$

Hence, it can be seen that for all n the operators $P_n R_1 P_n$ are invertible:[61]

$$(P_n R_1 P_n)^{-1} = (AM_-^{-1} + B\Lambda_+^{-1})P_n(M_+^{-1}A + \Lambda_-^{-1}B) \tag{2.6.33}$$

Moreover, $M_+^{\pm 1}(t), \Lambda_+^{\pm 1}(t), M_-^{\pm 1}(t)$, and $\Lambda_-^{\pm 1}(t)$ are in the spaces

$H_{\xi(\varepsilon)}^{+}(\Gamma) = AH_\varepsilon(\Gamma)$ and $H_{\xi(\varepsilon)}^{-}(\Gamma) = BH_\varepsilon(\Gamma) \oplus \{\text{constant}\}$ [26], respectively.

Thus, by using (2.6.33) and Lemma (2.6.2), we obtain:

$$\left\| (P_n R_1 P_n)^{-1} P_n \right\|_\gamma \leq c_{10} \ln^2 n \tag{2.6.34}$$

Furthermore, by following Muskhelishvili [24] for any function $h(t) \in H_\varepsilon(\Gamma)$ the operators BhA and AhB map $H_\gamma(\Gamma)$ into $H_{\xi(\varepsilon)}(\Gamma)$ and are continuous on $H_\gamma(\Gamma)$, $(0<\gamma<\varepsilon)$ and the same does the operator K. Hence, R_2 maps $H_\gamma(\Gamma)$ into $H_\xi(\varepsilon)(\Gamma)$ (R_2: $H_\gamma(\Gamma) \rightarrow H_{\xi(\varepsilon)}(\Gamma)$) and is continuous on $H_\gamma(\Gamma)$.

Beyond the above, let us assume the following relations:

$$y(t) = R_1^{-1} h(t) \tag{2.6.35}$$

$$y_n(t) = (P_n R_1 P_n)^{-1} P_n h(t) \tag{2.6.36}$$

and that $y_n^*(t)$ is the best uniform approximation of $y(t)$.

Therefore, since it is valid:

$$y_n(t) - y_n^*(t) = (P_n R_1 P_n)^{-1} P_n R_1 (y_n - y_n^*)(t) \tag{2.6.37}$$

by using (2.6.34), one obtains:

$$\begin{aligned} \|y - y_n\|_\gamma &\le \|y - y_n^*\|_\gamma + c_{10} \ln^2 n \|R_1\|_\gamma \|y_n - y_n^*\|_\gamma \le \\ &\le c_{11} H(y;\varepsilon) n^{\gamma - \xi(\varepsilon)} [1 + c_{10} \ln^2 n \|R_1\|_\gamma] \end{aligned} \tag{2.6.38}$$

Hence, from (2.6.38) we obtain the following inequality for the function $h(t) \in H_\varepsilon(\Gamma)$:

$$\left\| R_1^{-1} h - (P_n R_1 P_n)^{-1} P_n^h \right\|_\gamma \le c_{12} \ln^2 n \cdot n^{\gamma - \xi(\varepsilon)} H(R_1^{-1} h; \xi(\varepsilon)) \tag{2.6.39}$$

Thus, because of the invertibility of $R_1 + R_2$ and the continuity of R_2 we conclude that, for sufficiently large n, the operator $\Pi_n = P_n(R_1 + R_2)P_n$ is invertible and the following inequality is valid:

$$\left\| \Pi_n^{-1} \right\|_\gamma \le c_{13} \ln^2 n \tag{2.6.40}$$

and therefore the linear system (2.6.25) has a unique solution.

Finally, in order the requested inequality (2.6.27) to be proved the following relation will be used:

$$\|u - u_n\|_\gamma = \left\| (I - \Pi_n^{-1} P_n \Pi)(u - u_n^*) \right\|_\gamma \tag{2.6.41}$$

in which u_n^* is an arbitrary element of $P_n H_\gamma(\Gamma)$ and thus, the proof of Theorem 2.6.3 is completed.

A new technique has been proposed for the approximation of functions in a complex domain. These functions are defined in Banach spaces $H_\gamma(\Gamma)$, of functions satisfying Hölder's condition with exponent $\gamma (0 < \gamma \le 1)$, where Γ is a closed Lyapounov contour. Moreover, these results have been further used in order to prove the existence and uniqueness of solutions for the linear equation systems, on which the singular integral equations are reduced.

2.7. Numerical Recipes of Finite-Part Singular Integral Equations in Hilbert Spaces

By the present section the finite-part singular integral equations defined on Hilbert spaces are investigated, when their singularity consists of a homeomorphism of the integration interval, which is a unit circle, on itself. Therefore, some existence theorems are proposed for the solution of the above type of singular integral equations, approximated by several systems of linear algebraic equations [17].

Definition 2.7.1

Let the finite-part singular integral equation:

$$\Phi u(t) \equiv A(t)u(t) + \frac{B(t)}{\pi i} \oint_{\Gamma} \frac{u(x)}{(x-t)^{\mu}} \mathrm{d}x + C(t)u[\phi(t)] +$$

$$+ \frac{D(t)}{\pi \iota} \oint_{\Gamma} \frac{u(x)}{(x-\phi(t))^{\mu}} \mathrm{d}x + \int_{\Gamma} k(t,x)u(x)dx = f(t),\ t \in \Gamma, \mu \in N \tag{2.7.1}$$

in which Γ denotes the unit circle $\Gamma=\{t: |t|=1\}$, $A(t)$, $B(t)$, $C(t)$, $D(t)$, $k(t,x) \in H_{\beta, m\times m}$, $u(t)$ the unknown function and $\phi(t)$ a homeomorphism of Γ on itself.

Theorem 2.7.1

Consider the finite-part singular integral equation (2.11.1), in which $\phi(t)$ satisfies the following condition:

$$\phi_i(t) = t, \phi_j(t) = \phi[\phi_{j-1}(t)],\ 1 \le j \le 1,\ i \in N \tag{2.7.2}$$

An approximate solution of (2.7.1) is of the form:

$$u_n(t) = \sum_{l=-n}^{n} \varepsilon_L t^l,\ t = e^{is} \tag{2.7.3}$$

where the coefficients ε_l, l=-n,...n are obtained by solving the following system of linear algebraic equations:

$$\sum_{l=0}^{n} A_{j-l}\varepsilon_l + \sum_{l=-n}^{-1} \mathrm{B}_{j-l}\varepsilon_l + \sum_{l=0}^{n} C_{jl}\varepsilon_l + \sum_{l=-n}^{-1} D_{jl}\varepsilon_l + \sum_{l=-n}^{n} K_{jl}\varepsilon_l = f_j \tag{2.7.4}$$

(j=-n,...n)

with A_j, B_j, C_{jl}, D_{jl} and K_{jl}, j = ± 1, ± 2, ... , the Fourier coefficients of the corresponding matrix-valued functions:

$$A_1(t) = A(t) + B(t),\ B_1(t) = A(t) - B(t),\ C_{1l}(t) = [C(t) + D(t)][\phi(t)]^l$$

$$D_{1l}(t) = [C(t) - D(t)][\phi(t)]^l,\ K_{1l} = \int_\Gamma k(t,x)x^l dx \tag{2.7.5}$$

Beyond the above, consider the following conditions to be satisfied:

$$D(t) = C(t)A^{-1}[\phi(t)]B[\phi(t)] \tag{2.7.6}$$

$$\phi'(t) \neq 0,\ t \in \Gamma,\ \phi'(t) \in \mathrm{H}_\beta \tag{2.7.7}$$

and the operator Φ is invertible in the Hilbert space $L_{2,m}$ [5], det $A(t)$ does not vanish if $t \in \Gamma$ and $\phi(t)$ satisfies the following inequality:

$$\left\| C[\phi^{-1}(t)]A^{-1}(t)[\phi^{-1}(t)]^{1/2} \right\| < 1 \tag{2.7.8}$$

where $\phi^{-1}(t)$ is the inverse of $\phi(t)$.

Then, for sufficiently large n, the system (2.7.4) has a unique solution and the approximate solution (2.7.3) converges to the exact solution of (2.7.1) at a rate described by the following inequality:

$$\|u(t) - u_n(t)\|_{L_{2,m}} \leq \xi n^{-\beta} \tag{2.7.9}$$

in which ξ denotes a constant independent of n.

Proof. Consider the following operators to be valid on the space L_2:

$$\begin{aligned}
Fu(t) &= \frac{1}{\pi i} \oint_\Gamma \frac{u(x)}{(x-t)^\mu} \\
G &= \frac{1}{2}(I + F) \\
H &= \frac{1}{2}(I - F) \\
Ku(t) &= \int_\Gamma k(t,x)u(x)dx \\
M_n u(t) &= \sum_{i=-n}^{n} \varepsilon_i t^i \\
Lu(t) &= u[\phi(t)]
\end{aligned} \tag{2.7.10}$$

with ε_i the Fourier coefficients of $u(t)$ and I the identity operator.

Moreover, we introduce the following operator:

$$\Pi = AI + BF + Ku \tag{2.7.11}$$

where A and B are the operators corresponding to multiplication by $A(t)$ and $B(t)$, respectively.

Furthermore, the operator Π_n is on the subspace ImM_n of the space $L_{2,m}$ and is defined by the relation:

$$\Pi_n = M_n \Pi M_n \tag{2.7.12}$$

Thus, from (2.7.8) follows that the operator $\Pi_2 = I + CA_* L$ has an inverse, where A_* is the operation of multiplying by $A_*(t) = A^{-1}[\,\phi(t)\,]$. Therefore, from (2.7.6) one obtains:

$$\Phi = (I + CA_* L)(A_1 G + B_1 H + K_*) \tag{2.7.13}$$

where:

$$K_* = \Pi_2^{-1} K \tag{2.7.14}$$

As the operator Φ is invertible, then the following operator, is invertible, too:

$$\Phi_* = \Pi_2^{-1} \Phi = A_1 G + B_1 H + K_* \tag{2.7.15}$$

Furthermore, let us consider the following operators:

$$\Pi_{2n} = M_n \Pi_2 M_n = M_n (I + CA_* L) M_n = M_n + M_n CA_* L M_n \tag{2.7.16}$$

and:

$$\Phi_{2n} = (A_{1n} G + B_{1n} H + M_n K_*) M_n \tag{2.11.17}$$

in which A_{1n} and B_{1n} denote multiplication by the matrices of polynomials of degree not higher than n, uniformly approximating most accurately the matrices $A_1(t)$ and $B_1(t)$, respectively.

Beyond the above, the operators Π_{2n} are invertible for all n, and the operators Φ_{2n} are invertible for sufficiently large n.

By putting:

$$\varepsilon_n = \| M_n \Phi M_n - \Pi_{2n} \Phi_{2n} \| \tag{2.7.18}$$

we obtain

$$\lim_{n \to \infty} \varepsilon_n = 0$$

Hence, the operators $M_n \Phi M_n$ are invertible, beginning with some $n=n_1$ and therefore, the system (2.7.4) has a unique solution and the proof of the requested inequality (2.7.9) is obvious.

Theorem 2.7.2

Consider the finite-part singular integral equation (2.7.1), where $\varphi(t)$ satisfies condition (2.7.2). An approximate solution of (2.7.1) is of the form (2.7.3) where the coefficients ε_l, $l=-n,...n$ are obtained by solving the following system of linear algebraic equations:

$$\sum_{k=0}^{n}[A_{j-k} + \sum_{l=-n}^{n} D_{jl} A_{l-k}]\varepsilon_1 + \sum_{k=-n}^{-1}\left[B_{j-k} + \sum_{l=-n}^{n} D_{jl} B_{l-k}\right]\varepsilon_k +$$
$$\sum_{k=-n}^{n} K_{jl}\varepsilon_k = f_j\,,\ \ (j=\overline{-n,n}) \tag{2.7.19}$$

in which A_j, B_j, D_{jl} and K_{jl} , $j=\pm 1, \pm 2, ...$, are the Fourier coefficients of the matrix-valued functions given by (2.7.5).

Hence, if conditions (2.7.7) and (2.7.8) are satisfied, and the operator K is continuously invertible in the Hilbert space $L_{2,m}$ then the system (2.7.19) has a unique solution for sufficiently large n and the approximate solutions $u_n(t)$ converge to the exact solution $u(t)$ of the finite-part singular integral equation (2.7.1) with a rate given by (2.7.9).

Proof. If we use inequality (2.7.8), then the linear system (2.7.19) is equivalent to the following operational-equation system:

$$\Phi_n u_n \equiv M_n(I + CA_* L)M_n(AI + BF)M_n u_n +$$
$$M_n K M_n u_n = M_{nf} \tag{2.7.20}$$

where the operators M_n, L, F and K are given by (2.7.10), I denotes the identity operator and $A_*(t)=A^{-1}[\varphi(t)]$.

Beyond the above, the operator $AI+BF$ is invertible and because of the conditions of the theorem, beginning with some $n=n_*$, then the operators $Z_n=M_n(AI+BF)M_n$ are also invertible. Therefore, the operators $Z_n^{-1}M_n$ converge strongly to $(AI+BF)^{-1}$.

Furthermore, the operators Π_{2n} are invertible for all n and therefore, the operators $X_n = M_n(I+CA_*L)M_n(AI+BF)M_n$ are invertible for $n \geq n_*$ and the operators $X_n^{-1}M_n$ converge strongly to $(AI+BF)^{-1}(I+CA_*L)^{-1}$. Hence, the system (2.7.19) has a unique solution for sufficiently large n and the proof of Theorem 2.7.2 is completed.

This method is further used in order to prove, the existence of solutions for systems of finite-part singular integral equations, too. The singularity of the above systems consists of a system of diffeomorphisms of the integration interval, which is a unit circle, on itself.

Existence Theorems of Other Kinds of Finite-Part Singular Integral Computational Recipes in Hilbert Spaces

Definition 2.7.2

Let the finite-part singular integral equation:

$$Nu(t) \equiv A(t)u(t) + \frac{B(t)}{\pi i} \oint_{\Gamma} \frac{u(x)}{(x-t)^{\mu}} \mathrm{d}x + C(t)u[\phi(t)] +$$

$$+ \frac{D(t)}{\pi i} \oint_{\Gamma} \frac{u[\phi(\tau)]}{(x-t)^{\mu}} \mathrm{d}x + \int_{\Gamma} k(t,x)u(x)\mathrm{d}x = f(t), \ t \in \Gamma, \mu \in \mathrm{N} \qquad (2.7.21)$$

where Γ denotes the unit circle $\Gamma=\{t:|t|=1\}$, $A(t)$, $B(t)$, $C(t)$, $D(t)$, $k(x,t) \in H_{\beta,mxm}$, $f(t) \in H_{\beta,m}$, $u(t)$ the unknown function and $\phi(t)$ a homeomorphism of Γ on itself.

Theorem 2.7.3.

Consider the finite-part singular integral equation (2.7.21) , where an approximate solution is of the form (2.7.3) with the coefficients ε_l, $l=-n,...n$ to be obtained by solving a system of linear algebraic equations of the following form:

$$\sum_{k=0}^{n} A_{j-k}\varepsilon_k + \sum_{k=-n}^{-1} B_{j-k}\varepsilon_k + \sum_{l=-n}^{n} [C^*_{jl} + D^*_{jl} + K^*_{jl}]\varepsilon_l = f_j \qquad (2.7.22)$$

$$(j=-n,...n)$$

where A_j, B_j, and K_{jl} , $j=\pm 1, \pm 2, ...$ are the Fourier coefficients of the matrix-valued functions given by (2.7.5) and C^*_{jl} and D^*_{jl} are the Fourier coefficients of the matrix-valued functions $C(t)[\ \phi(t)\ ]^l$ and $\frac{D(t)}{\pi i}\int_{\Gamma}\frac{|\phi(t)|^l}{x-t}$, respectively.

Furthermore, if condition (2.7.7) is satisfied and the operator N is invertible in $L_{2,m}$, then the system (2.7.22) has unique solutions for sufficiently large n and the approximate solutions of (2.7.22) converge to the exact solution with a rate given by (2.7.9).

Proof. The following representation for the finite-part singular integral equation (2.7.21) will be used:

$$N = \Phi + FL - LF \qquad (2.7.23)$$

in which Φ denotes the finite-part singular integral equation (2.7.1) and F, L are the operators given by (2.7.10).

Thus, since the operator FL-LF , under the assumption concerning $\phi(t)$, is completely continuous it is easily proved that the system (2.7.22) has unique solutions for sufficiently large n.

Definition 2.7.3

Let the system of finite-part singular integral equations:

$$Tu(t) \equiv A(t)u(t) + \frac{B(t)}{\pi i} \oint_{\Gamma} \frac{u(x)}{(x-t)^{\mu}} \mathrm{d}x + \sum_{k=1}^{\xi} [C_k(t)u[\phi_k(t)] +$$

$$+ \frac{D_k(t)}{\pi i} \oint_{\Gamma} \frac{u(x)}{(x-\phi_k(t))^{\mu}}] + \int_{\Gamma} k(t,x)u(x)\mathrm{d}x = f(t) \tag{2.7.24}$$

where Γ denotes the unit circle $\Gamma=\{t:|t|=1\}$, $A(t)$, $B(t)$, $C_k(t)$, $D_k(t)$, $k(t,x) \in H_{\beta,mxm}$ $(0<\beta\leq 1)$, $f(t) \in H_{\beta,m}$, $u(t)$ the unknown function and ϕ_k, $k=1,2,...\xi$ a system of diffeomorphisms of Γ on itself.

The proof of the following theorem is analogous to the proof of Theorem 2.7.1:

Theorem 2.7.4.

Consider the system of finite-part singular integral equations (2.7.24), while an approximate solution is of the form (2.7.3) with the coefficients ε_l, $l=-n,...n$ to be obtained by solving a system of linear algebraic equations as following:

$$\sum_{l=0}^{n} A_{j-l}\varepsilon_l + \sum_{l=-n}^{-1} B_{j-l}\varepsilon_l + \sum_{k=1}^{\xi}\left[\sum_{l=0}^{n} C_{jl}^{(k)}\varepsilon_l + \sum_{l=-n}^{-1} D_{jl}^{(k)}\varepsilon_l\right] +$$

$$\sum_{l=-n}^{n} K_{jl}\varepsilon_l = fj,\ (j=-n,..n) \tag{2.7.25}$$

in which A_j, B_j and K_{jl} are the Fourier coefficients of the corresponding matrix-valued functions given by (2.7.5) and $C_{jl}^{(k)}$, $D_{jl}^{(k)}$ are the Fourier coefficients of $[C_k(t)+D_k(t)](\varphi_k(t))^l$ and $[C_k(t)-D_k(t)](\varphi_k(t))^l$, respectively.

Also, if condition (2.7.7) is satisfied and the operator T is invertible in $L_{2,m}$, then the system (2.7.25) has unique solutions for sufficiently large n.

Thus, a finite-part singular integral equations analysis has been presented, by proposing several approximation methods. Some existence theorems were proved, for the solutions of the systems of linear algebraic equations on which the finite-part singular integral equations are approximated. The singularity of the above type of singular integral equations consists of a homeomorphism of the integration interval (unit circle) on itself.

The proposed method was further extended in order to prove the existence of solutions for systems of finite-part singular integral equations, when their singularity consists of a system of diffeomorphisms of the integration interval (unit circle) on itself.

Hence, the present section was devoted to a basic description of numerical schemes, the vigorous foundation and comparison of a series of approximate methods and algorithms, and their application to the numerical solution of finite-part singular integral equations defined on Hilbert spaces.

REFERENCES

[1] J. Hadamard, Lectures on Cauchy's Problem in Linear Partial Differential Equations, Yale University Press, Yale (1932).
[2] J. Hadamard, Le Probléme de Cauchy et les Équations aux Derivées Partielles Linéaires Hyperboliques, Hermann, Paris (1932).
[3] L. Schwartz, Théorie des Distributions, Hermann, Paris (1966).
[4] H.R. Kutt, The numerical evaluation of principle value integrals, by finite-part integration, *Num. Math.* 24, 205-210 (1975).
[5] M.A. Golberg, The convergence of several algorithms for solving integral equations with finite-part integrals, *J. Int. Eq.* 5, 329-340 (1983).
[6] M.A. Golberg, The numerical solution of Cauchy singular integral equations with constant coefficients, *J. Int. Eq.* 9, 127-151 (1985).
[7] A.C. Kaya and F. Erdogan, On the solution of integral equations with strongly singular kernels, Q. *Appl. Math.* 45, 105-122 (1987).
[8] A.C. Kaya and F. Erdogan, On the solution of integral equations with a generalized Cauchy kernel, *Q. Appl. Math.* 45, 455-469 (1987).
[9] E.G. Ladopoulos, On the numerical solution of the finite-part singular integral equations of the first and the second kind used in fracture mechanics, *Comp. Meth. Appl. Mech. Engng* 65, 253-266 (1987).
[10] E.G. Ladopoulos, On the numerical evaluation of the general type of finite-part singular integrals and integral equations used in fracture mechanics, *J. Engng Fract. Mech.* 31, 315-337 (1988).
[11] E.G. Ladopoulos, The general type of finite-part singular integrals and integral equations with logarithmic singularities used in fracture mechanics, *Acta Mech.* 75, 275-285 (1988).
[12] E.G. Ladopoulos, Finite-part singular integro-differential equations arising in two-dimensional aerodynamics, *Arch. Mech.* 41, 925-936 (1989).
[13] E.G. Ladopoulos, Systems of finite-part singular integral equations in Lp applied to crack problems, *J. Engng Fract. Mech.* 48, 257-266 (1994).
[14] E.G. Ladopoulos, New aspects for the generalization of the Sokhotski-Plemelj formulae for the solution of finite-part singular integrals used in fracture mechanics, *Int. J. Fract.* 54, 317-328 (1992).
[15] E.G. Ladopoulos, V.A. Zisis and D. Kravvaritis, Singular integral equations in Hilbert space applied to crack problems, *Theor. Appl. Fract. Mech.* 9, 271-281 (1988).
[16] E.G. Ladopoulos, D. Kravvaritis and V.A. Zisis, Finite-part singular integral representation analysis in Lp of two-dimensional elasticity problems, *J. Engng Fract. Mech.* 43, 445-454 (1992).
[17] E.G. Ladopoulos, G. Tsamasphyros and V.A. Zisis, Finite-part singular integral approximations in Hilbert spaces, *Int. J. Math. Math. Scien.* 2004, 2787-2793 (2004).

[18] E.G. Ladopoulos and G. Tsamasphyros, Approximations of singular integral equations on Lyapounov contours in Banach spaces, *Comp. Math. Appl.* 50, 567-573 (2005).

[19] Harnack, Beiträge zur Theorie des Cauchy'schen Integral, Ber.d. k.6 Ges. Wiss., *Math. Phys.* Ch. 37, 379-398 (1885).

[20] D. Hilbert, Über eine Anwendung der Integralgleichungen auf ein Problem der Fuktionentheorie, Verhand. 3 I. Math. Kong., Heidelberg (1904).

[21] D. Hilbert, Grundzüge einer allgemeinen Theorie der linearen Integral-gleichungen, Leipzig (1912).

[22] H. Poincaré, Lecons de Méchanique Céleste, Vol. 3, Ch. 10, Gauthier et Villars, Paris (1910).

[23] J. Plemelj, Ein Ergänzungssatz zur Cauchyschen Integraldarstellung analytischer Funktionen, Randwerte betreffend, Monat. *Math. Phys.* 19, 205-210 (1908).

[24] F. Nöther, Über eine Klasse singulärer Integralgleichungen, *Math. Annal.* 82, 42-63 (1921).

[25] N.I. Muskhelishvili, *Some Basic Problems of the Mathematical Theory of Elasticity*, Noordhoff, Groningen, The Netherlands (1953).

[26] N.I. Muskhelishvili, *Singular Integral Equations,* Noordhoff, Groningen, The Netherlands (1972).

[27] N. Wiener and E. Hopf, Über eine Klasse singulärer Integralgleichungen, Sitz Berl. Akad. Wiss., 696-706 (1931).

[28] F.D. Gakhov, On the Rieman boundary value problem, *Matem. Sborn.* 2, 673-683 (1937).

[29] F.D. Gakhov, *Boundary Value Problems,* Pergamon Press, New York (1966).

[30] N.P. Vekua, *Systems of Singular Integral Equations*, Noordhoff, Groningen, The Netherlands (1967).

[31] V.V. Ivanov, The Theory of Approximate Methods and their Application to the Numerical *Solution of Singular Integral Equations*, Noordhoff, Leyden, The Netherlands (1976).

[32] W. Pogorzelski, *Integral Equations and their Applications*, Vol. 1, Pergamon and PWN-Polish Scientific Publ., Warszawa (1966).

[33] D. Elliot, Orthogonal polynomials associated with singular integral equations having a Cauchy kernel, *SIAM J. Math. Anal.* 13, 1041-1052 (1982).

[34] D. Elliot, The classical collocation method for singular integral equations having a Cauchy kernel, *SIAM J. Numer. Anal.* 19, 816-831 (1982).

[35] F.Erdogan, G.D. Gupta and T.S. Cook, Numerical solution of singular integral equations, in Methods of Analysis and Solutions of Crack Problems (Edited by G.C.Sih), Vol. 1, pp. 368-425, Noordhoff, Leyden, The Netherlands (1973).

[36] R.P. Gilbert and R. Magnanini, The boundary integral method for two-dimensional orthotropic materials, *J. Elasticity* 18, 61-82 (1987).

[37] D.A. Hills, D.N. Dai, P.A. Kelly and A.M. Korsunsky, *Singular Integral Equations in the Mechanics of Fracture*, Kluwer, Dordrecht (1995).

[38] G.C. Hsiao, P. Kopp and W.L. Wendland, A Galerkin collocation method for some integral equations of the first kind, *Computing* 25, 83-130 (1980).

[39] G.C. Hsiao, P. Kopp and W.L. Wendland, Some applications of a Galerkin-collocation method for integral equations of the first kind, *Math. Meth. Appl.* Sci. 6, 280-325 (1984).

[40] A.M. Korsunsky, Gauss-Chebyshev quadrature formulae for strongly singular integrals, *Q. Appl. Math.* 56, 461-472 (1998).
[41] E.G. Ladopoulos, On the solution of the two-dimensional problem of a plane crack of arbitrary shape in an anisotropic material, *J. Engng Fract. Mech.* 28, 187-195 (1987).
[42] E.G. Ladopoulos, On a new integration rule with the Gegenbauer polynomials for singular integral equations, used in theory of elasticity, *Ing. Archiv* 58, 35-46 (1988).
[43] V.A. Zisis and E.G. Ladopoulos, Singular integral approximations in Hilbert spaces for elastic stress analysis in a circular ring with curvilinear cracks, *Indus. Math.* 39, 113-134 (1989).
[44] E.G. Ladopoulos, Singular Integral Equations, Linear and Non-linear Theory and its Applications in Science and Engineering, Springer-Verlag, Berlin, New York (2000).
[45] G. Monegato, On the weights of certain quadratures for the numerical evaluation of Cauchy principal value integrals and their derivatives, *Numer. Math.* 50, 273-281 (1987).
[46] G. Monegato, Numerical evaluation of hypersingular integrals, *J. Comp. Appl. Math.* 50, 9-31 (1994).
[47] S. Prossdorf, On approximate methods for the solution of one-dimensional singular integral equations, *Appl. Anal.* 7, 259-270 (1977).
[48] R.P. Srivastav, Numerical solution of singular integral equations using Gauss type formulas - I. Quadrature and collocation on Chebyshev nodes, *IMA J. Numer. Anal.* 3, 305-318 (1983).
[49] A.N. Teong and D.L.Clements, A boundary integral equation method for the solution of a class of crack problems, *J. Elasticity* 17, 9-21 (1987).
[50] E. Venturino, Error bounds of the Galerkin method for singular integral equations of the second kind, *BIT* 25, 413-419 (1985).
[51] E. Venturino, Recent developments in the numerical solution of singular integral equations, *J. Math. Anal. Applic.* 115, 239-277 (1986).
[52] P.P. Zabreyko, Integral Equations-A Reference Text, Noordhoff, Leyden, The Netherlands (1975).
[53] U. Zastrow, Solution of the anisotropic elastostatical boundary value problems by singular integral equations, *Acta Mech.* 44, 59-71 (1982).
[54] U. Zastrow, Numerical plane stress analysis by integral equations based on the singularity method, *Solid Mech. Arch.* 10, 113-128 (1985).
[55] L.V. Kantorovich and V.I. Krylov, Approximate Methods of Higher Analysis, Gostekhizdat, Moscow (1952).
[56] A.H. Stroud and D. Secrest, Gaussian Quadrature Formulas, Prentice-Hall, Englewood Cliffs, New Jersey (1966).
[57] A. Erdélyi, W. Magnus, F. Oberhettinger and F. Tricomi, *Higher Transcendental Functions,* Vol.2, McGraw-Hill, New York (1953).
[58] M. Abramovitz and I.A. Stegun, Handbook of Mathematical Functions, Dover, New York (1965).
[59] K. Mitchell, Tables of the function $\int_0^z \frac{-\log(1-y)}{y} dy$, with an account of some properties of this and related functions, *Philos. Magaz.* 40, 355-368 (1949).

[60] P.K. Suetin, Series in Faber Polynomials, *Naukova Dumka*, Moscow (1984).

[61] I. Gohberg and I.A. Feldman, Convolution Equations and Projection Methods for their Solution, *AMS, Trans. Math. Monographs* 41, Providence (1974).

Chapter 3

ELASTICITY AND FRACTURE MECHANICS BY FINITE-PART SINGULAR INTEGRAL EQUATIONS

3.1. INTRODUCTION

Recently, an increasing importance and interest has been concentrated on the theory of finite-part singular integral equations applied in elasticity, structural analysis and fracture mechanics theory. By having in mind the implications for different problems of engineering mechanics and mathematical physics, some restrictions will be imposed upon the unknown and the given functions appearing in the systems of finite-part singular integral equations under consideration, or in the boundary conditions of the problems considered.

Several plane and antiplane elasticity problems, not possessing a closed-form solution, can be successfully solved by reduction to a finite-part singular integral equation (or a system of such equations) along their boundary. The difficulty which appears in these equations is that the integral, in which the unknown function appears, contains besides a regular part, one more part not defined in the ordinary sense, but in the finite-part sense. These finite-part singular integrals can be only by computational methods.

The first scientists who solved fracture mechanics problems, by using Cauchy singular integral equation methods, were H.F. Bückner [1], [2] and F. Erdogan [3]. Beyond the above, A.C. Kaya and F. Erdogan [4], [5] investigated several important problems of fracture mechanics, which are reduced to the solution of finite-part singular integral equations.

On the other hand, E.G. Ladopoulos [6] - [10] introduced several approximation methods for the solution of the finite-part singular integral equations of the first and the second kind and has applied them to the solution of several crack problems. Some further fracture mechanics problems were solved by using finite-part singular integral equation methods, by E.G. Ladopoulos, V.A. Zisis and D. Kravvaritis [11], [12].

Beyond the above, over the last years several elasticity problems were solved by some scientists by using singular integral equation methods. Among them we shall mention the following: F. Erdogan, G.D. Gupta and T.S. Cook [13], G.M.L. Gladwell and A.H. England [14], R.P. Gilbert and R. Magnanini [15], D.A. Hills, D.N. Dai, P.A. Kelly and A.M. Korsunsky [16], A.M Korsunsky [17], E.G. Ladopoulos et al. [18], [19], V.V. Panasyuk, M.P. Savruk and A.P. Datsyshin [20], A.N. Teong and D.L. Clements [21] and U. Zastrow [22], [23].

By the current chapter the following elasticity and fracture mechanics problems are solved: Non-symmetric cross-shaped crack in an infinite plane, symmetric cross-shaped crack in an infinite plane, elastic stress analysis in a circular ring with curvilinear cracks, cracks normal to a bimaterial interface, cracks parallel to the free boundary of an elastic isotropic semiplane, cracks parallel to the free boundary of an elastic isotropic semiplane, cracks parallel to the free boundary of an isotropic semi-infinite solid, cracks in a bimaterial infinite and isotropic solid under antiplane shear, straight cracks in an infinite, isotropic solid, periodic array of cracks along a straight line in an infinite and isotropic solid and periodic array of parallel straight cracks in an infinite and isotropic solid.

3.2. Fracture Mechanics of a Non-Symmetric Cross-Shaped Crack in an Infinite Plane

Consider the fracture mechanics problem of a non-symmetric cross-shaped crack in an infinite isotropic and elastic solid subjected to a constant stress p (see: Figure 3.2.1). This problem has been previously solved by I. N. Sneddon and S. C. Das by using the technique of integral transform [24]. In the present section the above fracture mechanics problem will be reduced to the solution of singular integral equations [11].

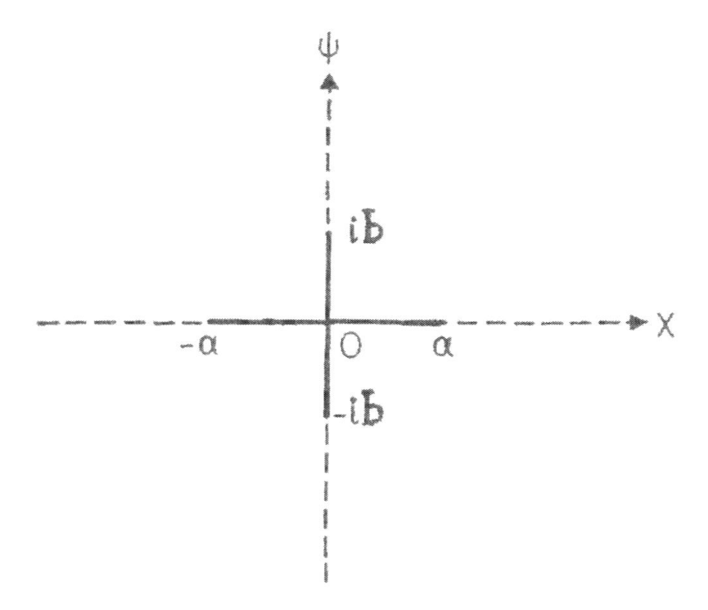

Figure 3.2.1. Nonsymmetric cross-shaped crack in an infinite medium.

The solution of this problem is given by the following system of singular integral equations of the first kind:

$$\int_0^1 \left[h_1(t)\frac{t}{t^2-x^2} + h_2(t)\frac{t\left[t^2-\left(\frac{a}{b}x\right)^2\right]}{\left[t^2+\left(\frac{a}{b}x\right)^2\right]^2} \right] \mathrm{d}t = -\frac{1}{2}\pi p$$
$$\int_0^1 \left[h_1(t)\frac{t\left[t^2-\left(\frac{b}{a}x\right)^2\right]}{\left[t^2+\left(\frac{b}{a}x\right)^2\right]^2} + h_2(t)\frac{t}{t^2-x^2} \right] \mathrm{d}t = -\frac{1}{2}\pi p \tag{3.2.1}$$

for $0 < x < 1$. The unknown functions h_i $(i = 1, 2)$ are equal to:

$$h_i(t) = -p\sqrt{\frac{t}{1-t}} f_i(t), \quad i = 1,2 \tag{3.2.2}$$

Based on the method of solution described earlier, then eqs. (3.2.1) can be numerically evaluated as following:

$$\sum_{m=1}^{n+1} C_m \left[f_1(t_m)\frac{t_m}{t_m^2-x_k^2} + f_2(t_m)\frac{t_m\left[t_m^2-\left(\frac{a}{b}x_k\right)^2\right]}{\left[t_m^2+\left(\frac{a}{b}x_k\right)^2\right]^2} \right] = \frac{1}{2}\pi$$
$$\sum_{m=1}^{n+1} C_m \left[f_1(t_m)\frac{t_m\left[t_m^2-\left(\frac{b}{a}x_k\right)^2\right]}{\left[t_m^2+\left(\frac{b}{a}x_k\right)^2\right]^2} + f_2(t_m)\frac{t_m}{t_m^2-x_k^2} \right] = \frac{1}{2}\pi \tag{3.2.3}$$

where $k = 1, 2,\ldots, n+1$. In eqs. (3.2.3), C_m, t_m and x_k are given by the formulas:

$$C_m = \frac{\pi}{n}\sin^2\frac{m\pi}{2n}, \quad m = 1,2,\ldots,n-1$$
$$t_m = \sin^2\frac{m\pi}{2n}, \quad m = 1,2,\ldots,n \tag{3.2.4}$$
$$x_k = \sin^2\frac{2k-1}{4n}\pi, \quad k = 1,2,\ldots,n$$

The stress intensity factors are of the forms:

$$K_1(a) = p\sqrt{2a}f_1(1), \qquad K_2(a) = p\sqrt{2b}f_2(1) \tag{3.2.5}$$

In Table 3.2.1 are given the numerical values of $f_j(1)$ (j = 1, 2). These results agree closely with $f_1(1) = 0.9802$ and $f_2(1) = 0.5897$ obtained in [24].

Table 3.2.1. Stress intensity factors for a non-symmetric cross-shaped crack

n		
2	0.979332	0.595234
3	0.981023	0.581867
4	0.980424	0.589342
5	0.980273	0.589668
6	0.980261	0.589669
7	0.980243	0.589671
8	0.980223	0.589676
9	0.980221	0.589680
10	0.980225	0.589683

3.3. Special Case of a Symmetric Cross-Shaped Crack in an Infinite Plane

By replacing $a = b$ in Figure 3.2.1, then the cross-shaped crack is said to be symmetric. This problem has been previously solved by D. P. Rooke and I. N. Sneddon [25] by using integral transforms and by A.Datsyshin and M. P. Savruck [26] by using a more general method. In the present chapter this fracture mechanics problem will be reduced to the solution of singular integral equations.

The governing equation is: [24]

$$\int_0^1 h(t)\left[\frac{t}{t^2 - x^2} + \frac{t(t^2 - x^2)}{(t^2 + x^2)^2}\right]\mathrm{d}t = -\frac{1}{2}\pi p, \quad 0 < x < 1 \tag{3.3.1}$$

By letting:

$$h(t) = -p\sqrt{\frac{t}{1-t}}f(t) \tag{3.3.2}$$

then eq. (3.3.1) reduces to:

$$\int_0^1 \sqrt{\frac{t}{1-t}}f(t)\left[\frac{t}{t^2 - x^2} + \frac{t(t^2 - x^2)}{(t^2 + x^2)^2}\right]\mathrm{d}t = \frac{1}{2}\pi, \quad 0 < x < 1 \tag{3.3.3}$$

The present method yields:

$$\sum_{m=1}^{n} C_m f(t_m)\left[\frac{t_m}{t_m^2 - x_k^2} + \frac{t_m(t_m^2 - x_k^2)}{(t_m^2 + x_m^2)^2}\right] = \frac{1}{2}\pi \tag{3.3.4}$$

in which C_m, t_m and x_k are given by eqs. (3.2.4). Because of symmetry, only K_1 prevails. Hence, Table 3.3.1 gives the numerical results that are in close agreement with $K_1/p\sqrt{2a} = 0.8636$ reported in [25].

Table 3.3.1. Stress intensity factor for a symmetric cross-shaped crack

n		n	
2	0.864137	6	0.863624
3	0.864272	7	0.863592
4	0.863826	8	0.863587
5	0.863752	9	0.863570
		10	0.863563

3.4. Fracture Behaviour of a Circular Ring with Curvilinear Cracks

Let a circular ring bounded by two circles Γ_1 and Γ_2 of radii R_1 and R_2 with the centers at the origin of the xOy-system (see Figure 3.4.1). Furthermore, the circular ring is weakened by a system of smooth, curvilinear cracks along the contours L_n ($n = 1, 2,\ldots, m$) referred to the local coordinate systems $x_nO_ny_n$.

The angles between the O_nx_n and the Ox axes are denoted by φ_n.

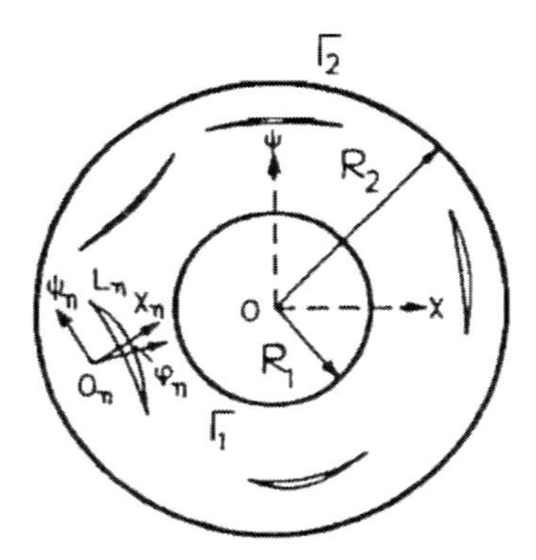

Figure 3.4.1. A circular ring weakened by a system of smooth, curvilinear cracks.

Beyond the above at the edges of the cracks act the following arbitrary self-equalibrating loads: [18]

$$M_n^{\pm} + T_n^{\pm} = p_n(t_n),\ t_n = z_n \in L_n,\ z_n = x_n + iy_n,\quad n = 1,2,\ldots,m \tag{3.4.1}$$

Muskhelishvili's [27] complex potentials $\Phi(z)$ and $\Psi(z)$ describing the state of stress in a circular ring with a free boundary can be written by the following relations:

$$\Phi(z) = \Phi_1(z) + \Phi_2(z) \tag{3.4.2}$$

$$\Psi(z) = \Psi_1(z) + \Psi_2(z) \tag{3.4.3}$$

where the complex potentials $\Phi_1(z)$ and $\Psi_1(z)$, which define the state of stress in an infinite plane caused by the displacement discontinuities $f_k(z_k)$ on the contours L_k ($k = 1, 2,\ldots, m$), are:

$$\Phi_1(z) = \frac{1}{2\pi}\sum_{k=1}^{m} e^{i\varphi_k} \int_{L_k} \frac{f_k'(t)\,\mathrm{d}t}{T_k - z} \tag{3.4.4}$$

with:

$$T_k = te^{i\varphi_k} + z_k^0, \quad z_k^0 = x_k^0 + iy_k^0 \tag{3.4.5-6}$$

and:

$$\Psi_1(z) = \frac{1}{2\pi}\sum_{k=1}^{m} \int_{L_k} \left[\frac{\overline{f_k'(t)}\,\overline{\mathrm{d}t}}{T_k - z} e^{-i\varphi_k} - \frac{\overline{T_k} f_k'(t)\,\mathrm{d}t}{(T_k - z)^2} e^{i\varphi_k} \right] \tag{3.4.7}$$

in which z_k^0 denote the complex coordinates with the origins O_k of the local coordinate system in the fundamental system.

Moreover, in (3.4.2) and (3.4.3), the complex potentials $\Phi_2(z)$ and $\Psi_2(z)$ are defined as:

$$\Phi_2(z) = \sum_{r=-\infty}^{\infty} c_r z^r, \quad \Psi_2(z) = \sum_{r=-\infty}^{\infty} b_r z^r \tag{3.4.8-9}$$

in which c_r and b_r are written as:

$$c_r = \frac{1}{2\pi(R_2 - R_1)^r} \sum_{k=1}^{m} \int_{L_k} \left[c_{rk}^1(t) f_k'\,\mathrm{d}t + c_{rk}^2(t)\overline{f_k'}(t)\overline{\mathrm{d}t} \right] \tag{3.4.10}$$

$$b_r = \frac{1}{2\pi(R_2 - R_1)^r} \sum_{k=1}^{m} \int_{L_k} \left[b_{rk}^1(t) f_k'\,\mathrm{d}t + b_{rk}^2(t)\overline{f_k'}(t)\overline{\mathrm{d}t} \right] \tag{3.4.11}$$

$r = 0, \pm 1, \pm 2,\ldots$

where:

$$c_{0k}^{i}(t)=\frac{1}{2[1-(R_1/R_2)^2]}\left[F_{0ki}^{2}(t)-(R_1/R_2)^2F_{0ki}^{1}(t)\right] \tag{3.4.12}$$

$$c_{1k}^{i}(t)=\frac{1}{R_2[1-(R_1/R_2)^4]}\left[\overline{F_{1kv}^{2}(t)}-(R_1/R_2)^3\overline{F_{1ki}^{1}(t)}\right] \tag{3.4.13}$$

$$c_{-1k}^{i}(t)=0 \tag{3.4.14}$$

$$c_{rk}^{i}(t)=(R_2-R_1)^{-r}\begin{bmatrix}\delta_{1r}F_{rki}^{1}(t)+\delta_{2r}F_{rki}^{2}(t)+\delta_{3r}\overline{F_{-rkv}^{1}(t)}+\\ \delta_{4r}\overline{F_{-rkv}^{2}(t)}\end{bmatrix} \tag{3.4.15}$$

$$r=\pm 2,\pm 3,\ldots$$

$$b_{rk}^{i}(t)=\left[\frac{R_1}{R_2-R_1}\right]^{-2(r+1)}c_{-(r+2)k}^{v}(t)-(1+r)\left[\frac{R_1}{R_2-R_1}\right]^{2}c_{(r+2)k}^{i}(t)$$
$$-\left[\frac{R_1}{R_2-R_1}\right]^{-r}F_{(r+2)kv}^{1}(t),\qquad r=0,\pm1,\pm2,\ldots,\quad i,v=1,2,\ \ i\neq v \tag{3.4.16}$$

$$F_{rkv}^{i}(t)=-\frac{1}{2\pi}\int_{0}^{2\pi}H_{vk}^{i}(t,e^{i\xi})e^{-ir\xi}\,\mathrm{d}\xi \tag{3.4.17}$$
$$r=0,\pm1,\pm2,\ldots,i,v=1,2$$

$$H_{1k}^{i}(t,e^{i\xi})=e^{-i\varphi_k}\left[\frac{1}{T_k-R_ie^{-i\xi}}-\frac{e^{2i\xi}(R_je^{-i\xi}-\overline{T_k})}{(T_k-R_je^{i\xi})^2}\right],\quad i=1,2 \tag{3.4.18}$$

$$H_{2k}^{i}(t,e^{i\xi})=e^{-i\varphi_k}\left[\frac{1}{\overline{T_k}-R_ie^{-i\xi}}-\frac{e^{2i\xi}}{T_k-R_je^{i\xi}}\right],\quad i=1,2 \tag{3.4.19}$$

$$\delta_r^1=-(R_1/R_2)^{2-r}\delta_r^2,\quad \delta_r^2=(1+r)[1-(R_1/R_2)^2]/\delta_0 \tag{3.4.20-21}$$

$$\delta_r^3=(R_1/R_2)^{2+r}\delta_r^4,\quad \delta_r^4=-[1-(R_1/R_2)^{-2r+2}]/\delta_0 \tag{3.4.22-23}$$

$$\delta_0=\begin{Bmatrix}(1-r^2)[1-(R_1/R_2)^2]^2-\\ -[1-(R_1/R_2)^{2r+2}][1-(R_1/R_2)^{-2r-2}\end{Bmatrix}\left[\frac{R_2}{R_2-R_1}\right]^{r} \tag{3.4.24}$$

By using relations (3.4.2) to (3.4.24) to satisfy the boundary conditions (3.4.1), then the following system of m singular integral equations is obtained:

$$\sum_{k=1}^{n}\int_{L_k}\left[K_{nk}(t,t')f_k{}'(t)\mathrm{d}t+R_{nk}(t,t')\overline{f_k'(t)\mathrm{d}t}\right]=\pi p_n(t_n{}') \tag{3.4.25}$$
$$t_n{}'\in L_n,n=1,2,\ldots,m$$

with the boundary conditions:

$$\int_{L_n} f_n'(t)\,\mathrm{d}t = 0, \quad n = 1,2,\ldots,m \tag{3.4.26}$$

where:

$$\begin{aligned} K_{nk}(t,t') &= \frac{1}{2}e^{i\varphi_k}\left[\frac{1}{T_k - T_n'} + \frac{e^{-2i\varphi_n}}{\overline{T}_k - \overline{T}'_n}\cdot\frac{\overline{\mathrm{d}t'_n}}{\mathrm{d}t'_n}\right] \\ &+\frac{1}{2}\sum_{r=-\infty}^{\infty}\left[\begin{array}{l} c_{rk}^1(t)T_n'^r + c_{rk}^2(t)\overline{T}'^r_n + e^{-2i\varphi_n}\dfrac{\overline{\mathrm{d}t'_n}}{\mathrm{d}t'_n} \\ \left\{r\overline{a_{rk}^2(t)T_n'}\,\overline{T}'^{r-1}_n + \overline{b_{rk}^2(t)}\overline{T}'^r_n\right\}\end{array}\right] \end{aligned} \tag{3.4.27}$$

$$\begin{aligned} R_{nk}(t,t') &= \frac{1}{2}e^{-i\varphi_k}\left[\frac{1}{\overline{T}_k - \overline{T}'_n} - \frac{(T_k - T_n')e^{-2i\varphi_n}}{(\overline{T}_k - \overline{T}'_n)^2}\cdot\frac{\overline{\mathrm{d}t'_n}}{\mathrm{d}t'_n}\right] \\ &+\frac{1}{2}\sum_{r=-\infty}^{\infty}\left[\begin{array}{l} c_{rk}^1(t)\overline{T}'^r_n + c_{rk}^2(t)T_n'^r + e^{-2i\varphi_n}\dfrac{\overline{\mathrm{d}t'_n}}{\mathrm{d}t'_n} \\ \left\{rc_{rk}^1(t)T_n'\overline{T}'^{r-1}_n + \overline{b}^1_{rk}(t)\overline{T}'^r_n\right\}\end{array}\right] \end{aligned} \tag{3.4.28}$$

with:

$$T_n' = t'e^{i\varphi_n} + z_n^0 \tag{3.4.29}$$

Hence, by solving the system of m singular integral equations (3.4.25) with the boundary conditions (3.4.26), then the displacement discontinuities $f_k(t)$ are determined. The above system is numerically evaluated by means of the Gauss-Jacobi quadrature formula described in Chapter 2.

Application to the Determination of the Elastic Behavior of a Circular Ring with a Rectilinear Crack Placed along the Ox-axis

As an application of the previous analysis, will be determined the elastic stress state in a circular ring with a rectilinear crack of length $2a$, placed along the Ox-axis, under applied pressure p. The center of the crack is situated at a distance d from the inner boundary of the ring (see: Figure 3.4.2). For the present case, $d = (R_2\text{-}R_1)/2$.

Therefore, for the above application, the following relations must be replaced into equations (3.4.25) to (3.4.29):

$$n = k = 1, \quad \varphi_1 = 0, \quad \overline{dt_1}/dt_1 = 0, \quad z_1^0 = R_1 + d \tag{3.4.30}$$

Thus, the system of m singular integral equations (3.4.25) reduces to the solution of one singular integral equation. For the numerical evaluation of this equation, the Gauss-Jacobi quadrature formula is used, which was described in Chapter 2.

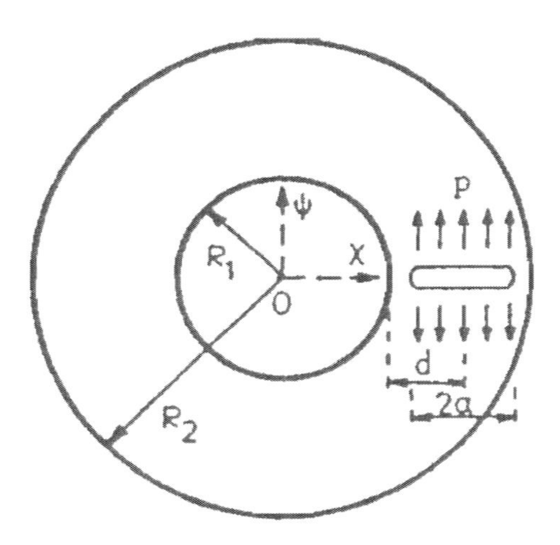

Figure 3.4.2. A circular ring weakened by a rectilinear crack of length $2a$, placed along the Ox-axis, under applied pressure p.

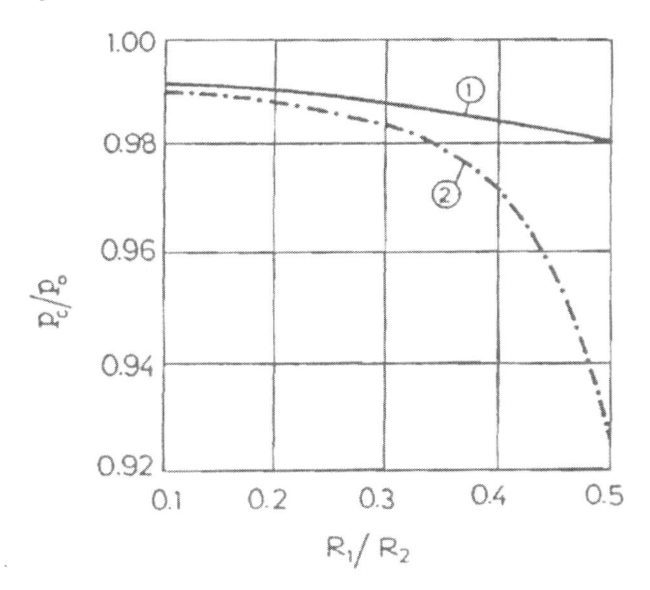

Figure 3.4.3. Relation between the ratios p_c / p_0 and R_1 / R_2 where line1 refers to the left end of the crack and line 2 to the right end.

Figure 3.4.3 shows the relation between the ratio p_c/p_0, where p_c denotes the critical load applied to the crack edges and p_0 the critical load acting at the crack in an infinite plane, and the ratio R_1/R_2. Line 1 refers to the left end of the crack and line 2 to the right end.

Beyond the above, as is easily seen from Figure 3.4.3, the critical load p_c is changed very little for the left end of the crack when the ratio R_1/R_2 varies.

On the other hand, for the right end of the crack, p_c decreases with increasing R_1/R_2, when $R_1/R_2 > 0.35$, while for $R_1/R_2 < 0.35$ the value of p_c for a circular ring is practically equal to that for an infinite plane weakened by a circular hole and crack.

3.5. Non-Dimensional Stress Intensity Factors near Cracks Normal to a Bimaterial Interface

As another application of the theory of finite-part singular integral equations, let us consider a straight crack of length $2a$, lying across the x-axis in an orthotropic half plane of a bimaterial interface consisting of isotropic and orthotropic materials, under constant strain to the interface applied at infinity (see Figure 3.5.1).

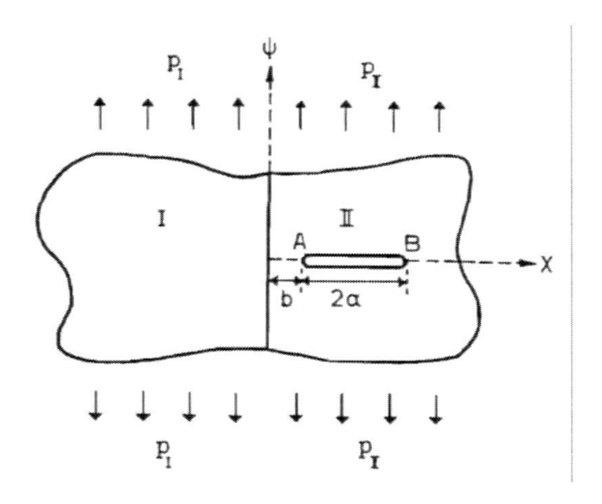

Figure 3.5.1. A straight crack in a bimaterial interface, consisting of isotropic (I) and orthotropic (II) solids, under applied pressure.

This problem is reduced to the solution of the singular integral equation: [9]

$$\frac{1}{\pi}\int_{-1}^{1}\frac{g(t)}{t-x}\,\mathrm{d}t+\frac{1}{\pi}\int_{-1}^{1}k(x,t)g(t)\,\mathrm{d}t=1,\qquad -1<x<1 \tag{3.5.1}$$

under the boundary condition:

$$\int_{-1}^{1}g(t)\,\mathrm{d}t=0 \tag{3.5.2}$$

in which the kernel $k(x,t)$ is valid as:

$$k(x,t) = \frac{m_0}{[t + x + 2(1+\zeta)]} + \frac{m_1}{[\mu_2^{1/2} t + \mu_1^{1/2} x + (\mu_1^{1/2} + \mu_2^{1/2})(1+\zeta)]} + \frac{m_2}{[\mu_1^{1/2} t + \mu_2^{1/2} x + (\mu_1^{1/2} + \mu_2^{1/2})(1+\zeta)]} \tag{3.5.3}$$

with:

$$\zeta = b/a \tag{3.5.4}$$

Beyond the above, in (3.5.3), m_0, m_1 and m_2 are constants depending on the elastic constants of the isotropic and orthotropic half plane, and μ_1 and μ_2 are the roots of the characteristic equation for the orthotropic solid [28], [29]:

$$c_{11}c_{33}\mu^2 + [c_{12}(2c_{33} + c_{12}) - c_{11}c_{33}]\mu + c_{22}c_{33} = 0 \tag{3.5.5}$$

where the elastic constants are:

$$c_{11} = E_x/(1 - v_{xy}v_{yx}) \tag{3.5.6}$$

$$c_{12} = v_{yx}E_x/(1 - v_{xy}v_{yx}) \tag{3.5.7}$$

$$c_{22} = E_y/(1 - v_{xy}v_{yx}) \tag{3.5.8}$$

$$c_{33} = G_{xy} \tag{3.5.9}$$

with E_x and v_{xy} the Young's modulus and Poisson's ratio in the plane of isotropy, E_y and v_{yx} the same quantities in the transverse direction and G_{xy} the shear modulus.

Generally, the solution of the singular integral equation (3.5.1) is obtained as:

$$g(t) = w(t)f(t) \tag{3.5.10}$$

in which the weight function $w(t)$ is valid:

$$w(t) = (1 - t^2)^{-1/2} \tag{3.5.11}$$

As the strain is constant to the interface applied at infinity, then the pressures p_{I} and p_{II} are defined as:

$$p_I/E = p_{II}/E_y$$

in which E denotes the Young's modulus of the isotropic solid.

Moreover, let us define the stress intensity factors K_A and K_B, at the points A and B of the crack (see: Figure 3.5.1):

$$K_A = \lim \sqrt{[-2\pi a(x+1)](\sigma_y^{II})_{y=0}} \tag{3.5.12}$$

$$K_B = \lim \sqrt{[2\pi a(x-1)](\sigma_y^{II})_{y=0}} \tag{3.5.13}$$

where:

$$(\sigma_y^{II})_{y=0} / p_{II} = -1/\pi \int_{-1}^{1} \frac{g(t)}{t-x} \mathrm{d}t, \ (t \le 1, -1-\zeta < t \le -1) \tag{3.5.14}$$

Hence, by using (3.5.10) and (3.5.11), then eq. (3.5.14) can be expressed by the following form:

$$(\sigma_y^{II})_{y=0} / p_{II} = \begin{cases} f(1)/\sqrt{(x^2-1)}, & x \ge 1 \\ -f(-1)/\sqrt{(x^2-1)}, & x \le -1 \end{cases} \tag{3.5.15}$$

By replacing (3.5.15) in (3.5.12) and (3.5.13) , one obtains:

$$K_A / p_{II} \sqrt{(\pi a)} = -f(-1) \tag{3.5.16}$$

$$K_B / p_{II} \sqrt{(\pi a)} = f(1) \tag{3.5.17}$$

Furthermore, by using the Lobatto-Chebyshev rule [9] for the numerical evaluation of the singular integral equation (3.5.1) we obtain $f(1)$ and $f(-1)$ directly.

As an application of the previous described method we consider two isotropic solids, either epoxy resin (ER) with $E = 3.13$ GPa, G=1.16 GPa and $v = 0.35$ or aluminium (Al) with $E = 69.92$ GPa, $G = 26.78$ GPa and $v = 0.3$, and two orthotropic solids, either a first kind of carbon fiber (CF_1) with $E_x = 8.62$ GPa, $E_y = 400$ GPa, $G_{xy}= 2.8$ GPa and $v_{yx} = 0.35$ or a second kind of carbon fiber (CF_2) with $E_x = 15$ GPa, $E_y = 232$ GPa, $G_{xy} = 5.02$ GPa and $v_{yx} = 0.28$.

Figure 3.5.2 shows the relation of the stress intensity factors $K_A / p_{II} \sqrt{(\pi a)}$ and $K_B / p_{II} \sqrt{(\pi a)}$ for the points A and B as a function of distance from the interface for CF_1/Al and CF_2/Al. Also, Fig. 3.5.3 shows the relation of the stress intensity factors $K_A / p_{II} \sqrt{(\pi a)}$ and $K_B / p_{II} \sqrt{(\pi a)}$ for the points A and B as a function of distance from the interface for CF_1/ER and CF_2/ER.

Finally, from Figures 3.5.2 and 3.5.3 it can be seen that as one of the crack tips in the fiber approaches the interface, the non-dimensional stress intensity factors at both ends increase when the matrix is an epoxy resin and decrease when it is aluminium.

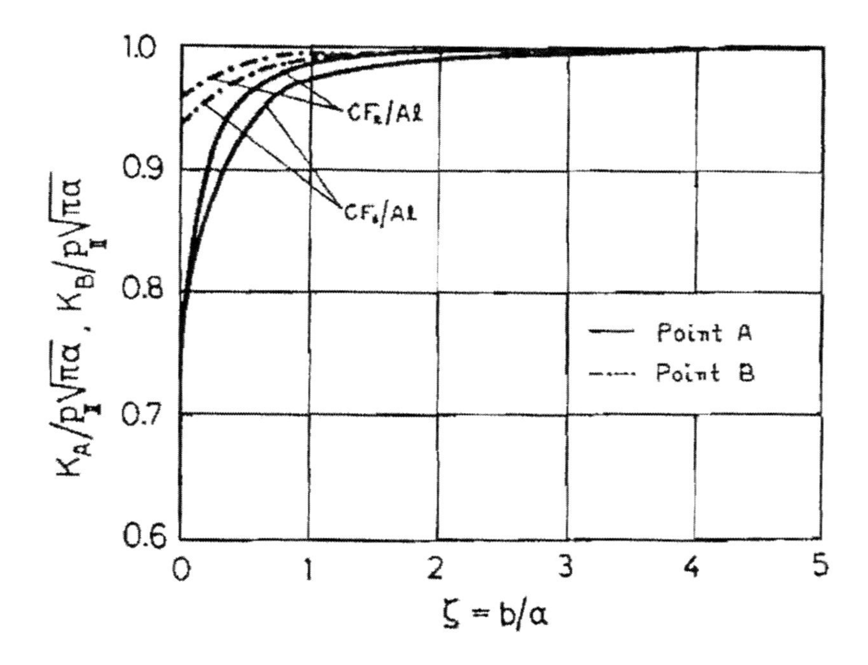

Figure 3.5.2. Relation of the stress intensity factors K_A/p_{II} [ ] and $K_B/p_{II}\ \sqrt{\pi a}$ for the points A and B, as a function of distance from the interface, for $CF_1\ /Al$ and $CF_2\ /Al$.

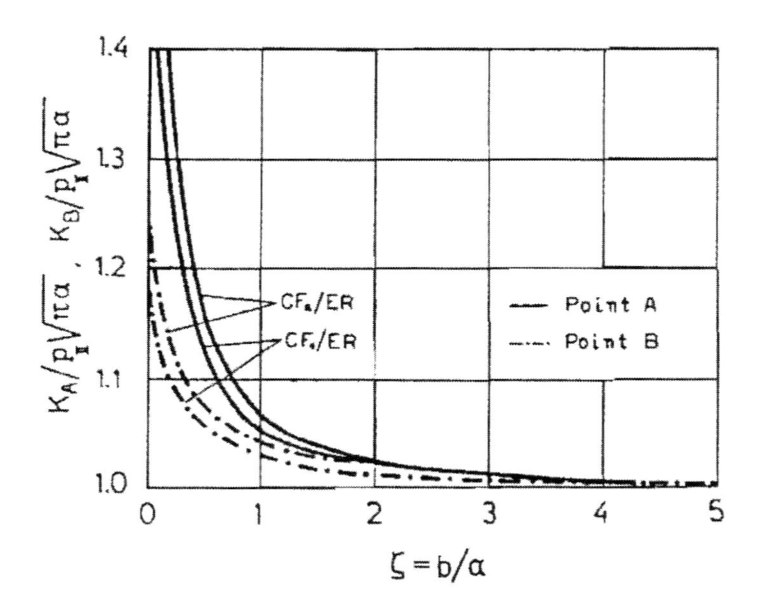

Figure 3.5.3. Relation of the stress intensity factors $K_A/p_{II}\ \sqrt{\pi a}$ and $K_B/p_{II}\ \sqrt{\pi a}$ for the points A and B, as a function of distance from the interface, for CF_1/ER and CF_2/ER.

3.6 Fracture Mechanics of Cracks Parallel to the Free Boundary of an Elastic Isotropic Semiplane

Let a crack of length a, parallel to the free boundary of an elastic isotropic semiplane (see Figure 3.6.1). A uniform pressure p applied to the crack surfaces is further assumed. [6]

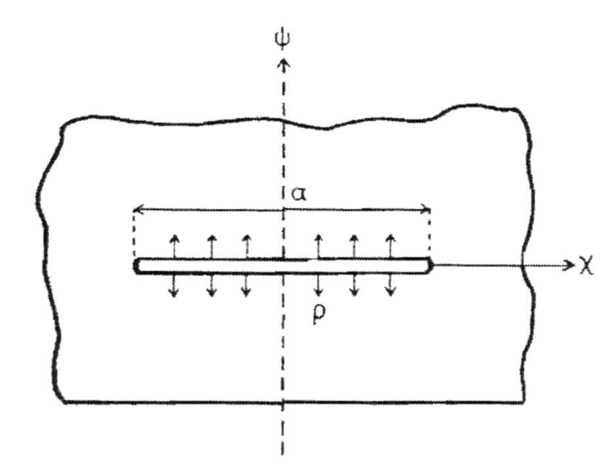

Figure 3.6.1. Crack parallel to the free boundary of an elastic isotropic semiplane.

The problem of determining the stress field in the neighbourhood of such a crack is reduced to the solution of the following singular integral equation on the integration interval [0,1], where $t = 0$ corresponds to the intersection point of the crack with the boundary, and $t =$ 1 to the crack peak [20]:

$$\int_0^1 h(t)\left[\frac{1}{t-x}+\frac{x^2-t^2+4tx}{(t+x)^3}\right]\mathrm{d}t=-\pi p,\quad 0<x<1 \tag{3.6.1}$$

Moreover, by representing the unknown function $h(t)$ in the form:

$$h(t)=-p\left(\frac{t}{1-t}\right)^{1/2}f(t) \tag{3.6.2}$$

then (3.6.1) reduces to:

$$\int_0^1\left(\frac{t}{1-t}\right)^{1/2}f(t)\left[\frac{1}{t-x}+\frac{x^2-t^2+4tx}{(t+x)^3}\right]\mathrm{d}t=\pi \tag{3.6.3}$$

Let us now use the following semi-open quadrature equations in order to evaluate numerically (3.6.3):

$$\sum_{m=1}^{n+1} C_m f(t_m)\left[\frac{1}{t_m - x_K} + \frac{x_K^2 - t_m^2 + 4t_m x_K}{(t_m + x_K)^3}\right] = \pi \tag{3.6.4}$$

where:

$$\begin{aligned} C_m &= \frac{\pi}{n}\sin^2\frac{m\pi}{2n}, \quad m = 1,2,\ldots,n-1 \\ t_m &= \sin^2\frac{m\pi}{2n}, \quad m = 1,2,\ldots,n \\ x_K &= \sin^2\frac{2K-1}{4n}\pi, \quad K = 1,2,\ldots,n \end{aligned} \tag{3.6.5}$$

On the other hand, by using the quadrature equations of the open type we further have:

$$\sum_{m=1}^{n} C_{m_1} f(t_{m_1})\left[\frac{1}{t_{m_1} - x_{K_1}} + \frac{x_{K_1}^2 - t_{m_1}^2 + 4t_{m_1} x_{K_1}}{(t_{m_1} + x_{K_1})^3}\right] = \pi \tag{3.6.6}$$

where:

$$\left.\begin{aligned} C_{m_1} &= \frac{2\pi}{2n+1}\sin^2\frac{\pi m}{2n+1} \\ t_{m_1} &= \frac{1}{2}\left(1-\cos\frac{2m\pi}{2n+1}\right) \\ x_{K_1} &= \frac{1}{2}\left(1-\cos\frac{2K-1}{2n+1}\pi\right) \end{aligned}\right\} \quad K,m = 1,2,\ldots,n \tag{3.6.7}$$

Beyond the above, the stress intensity factor at peaks of the crack has the following form:

$$K_1 = p\sqrt{a}f(1) \tag{3.6.8}$$

in which the value of $f(1)$ is determined by the system of (3.6.4), as following:

$$f(1) = \frac{2(-1)^n}{2n+1}\sum_{m=1}^{n}\frac{(-1)^m f(t_m)t_m}{\cos m\pi/(2n+1)} \tag{3.6.9}$$

Table 3.6.1 shows the values of the stress intensity factors by using either (3.6.4) or (3.6.6).

Finally, as it can be seen from Table 3.6.1, the numerical method converges identically to the exact value of the dimensionless stress intensity factor 1.1215 [20].

Table 3.6.1. Numerical evaluation of $K_1 = p\sqrt{a}$

n	(3.6.4)	(3.6.6)
10	1.123697	1.118012
20	1.122119	1.120543
30	1.121816	1.121029
40	1.121667	1.121468
50	1.121485	1.121213

3.7. Fracture Behaviour of Cracks Parallel to the Free Boundary of Isotropic Semi-Infinite Solids

Consider further the problem of a crack parallel to the free boundary of an isotropic semi-infinite plane, at a distance d from its free boundary. A uniform pressure p is applied to the crack surfaces (Figure 3.7.1).

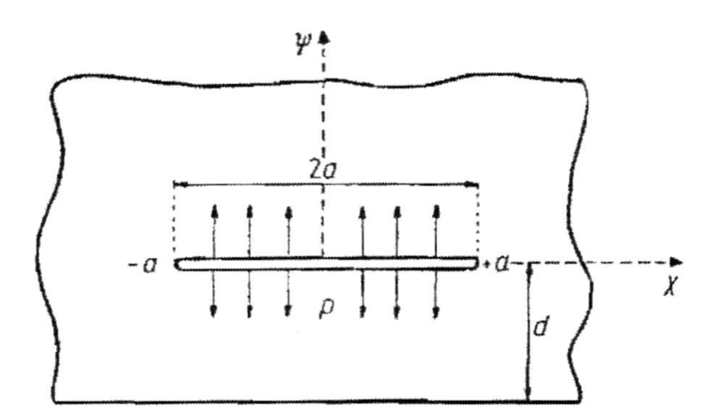

Figure 3.7.1. Crack parallel to the free boundary of an isotropic semi-infinite solid.

The following system of singular integral equations gives the solution of the above described problem, where the length of the crack is assumed to be $2a = 2$, cp [13], [20], [30]:

$$\int_{-1}^{1} \frac{g_1(t)\,\mathrm{d}t}{t-x} + \int_{-1}^{1} [K_{11}(x,t)g_1(t) + K_{12}(x,t)g_2(t)]\,\mathrm{d}t = 0$$

$$\int_{-1}^{1} \frac{g_2(t)\,\mathrm{d}t}{t-x} + \int_{-1}^{1} [K_{21}(x,t)g_1(t) + K_{22}(x,t)g_2(t)]\,\mathrm{d}t = -\pi p \frac{1+k}{2G} \qquad (3.7.1)$$

$$(-1 < k < 1)$$

in which G is the shear modulus and the values of K_{ij} $(i, j = 1, 2)$ are given by:

$$K_{11}(x,t) = -\frac{t-x}{(t-x)^2+4d^2} + \frac{8d^2(t-x)}{[(t-x)^2+4d^2]^2} - \frac{4d^2(t-x)[12d^2-(t-x)^2]}{[(t-x)^2+4d^2]^3}$$

$$K_{12}(x,t) = K_{21}(x,t) = -\frac{8d^3[4d^2-3(t-x)^2]}{[(t-x)^2+4d^2]^3} \tag{3.7.2}$$

$$K_{22}(x,t) = -\frac{t-x}{(t-x)^2+4d^2} - \frac{8d^2(t-x)}{[(t-x)^2+4d^2]^2} - \frac{4d^2(t-x)[12d^2-(t-x)^2]}{[(t-x)^2+4d^2]^3}$$

The values of $g_i(x)$ $(i = 1, 2)$ are given by the relations:

$$\begin{aligned} g_1(x) &= \frac{\vartheta}{\vartheta x}\left[u_1(x,+0) - u_1(x,-0)\right] \\ g_2(x) &= \frac{\vartheta}{\vartheta x}\left[u_2(x,+0) - u_2(x,-0)\right] \end{aligned} \tag{3.7.3}$$

where u_1 and u_2 are the displacements parallel and perpendicular to the cut, respectively. As the functions g_i have integrable singularities at the end points -1 and 1, then the constant in (3.7.1) is k=1. Furthermore, the following conditions are valid:

$$\int_{-1}^{1} g_i(x)\,\mathrm{d}x = 0 \tag{3.7.4}$$

The weight function for (3.7.1) is $w(t) = (1-t^2)^{-1/2}$ which is a special case of the Gegenbauer polynomials.

The stress intensity factors K_1 and K_2 are given as:

$$\begin{aligned} K_1 &= -\frac{2G}{1+k}\lim_{x\to 1}(1-x^2)^{1/2} g_2(x) \\ K_2 &= -\frac{2G}{1+k}\lim_{x\to 1}(1-x^2)^{1/2} g_1(x) \end{aligned} \tag{3.7.5}$$

where:

$$\lim_{x \to 1}(1-x^2)^{1/2} g_2(x) = \sum_{n=1}^{\infty} E_n$$
$$\lim_{x \to 1}(1-x^2)^{1/2} g_1(x) = \sum_{n=1}^{\infty} D_n \tag{3.7.6}$$

Also, the values of E_n and D_n are given by the system:

$$\frac{\pi}{2} E_K + \sum_{n=1}^{N} (E_{Kn}^{I} E_n + D_{Kn}^{I} D_n) = 0$$
$$\frac{\pi}{2} D_K + \sum_{n=1}^{N} (E_{Kn}^{II} E_n + D_{Kn}^{II} D_n) = 0, \qquad (K = 1,\ldots,N) \tag{3.7.7}$$

in which:

$$E_{Kn}^{I} = \int_{-1}^{1} C_{2K-1}(x) M_n^{11}(x)(1-x^2)^{1/2}\,\mathrm{d}x$$
$$D_{Kn}^{I} = \int_{-1}^{1} C_{2K-1}(x) M_n^{12}(x)(1-x^2)^{1/2}\,\mathrm{d}x \tag{3.7.8}$$
$$E_{Kn}^{II} = \int_{-1}^{1} C_{2K-2}(x) M_n^{21}(x)(1-x^2)^{1/2}\,\mathrm{d}x$$
$$D_{Kn}^{II} = \int_{-1}^{1} C_{2K-2}(x) M_n^{22}(x)(1-x^2)^{1/2}\,\mathrm{d}x$$

and:

$$M_n^{11}(x) = \frac{1}{\pi}\int_{-1}^{1} K_{11}(x,t) C_{2n}(t)(1-t^2)^{-1/2}\,\mathrm{d}t$$
$$M_n^{12}(x) = \frac{1}{\pi}\int_{-1}^{1} K_{12}(x,t) C_{2n-1}(t)(1-t^2)^{-1/2}\,\mathrm{d}t \tag{3.7.9}$$

$$M_n^{21}(x) = \frac{1}{\pi}\int_{-1}^{1} K_{21}(x,t) C_{2n}(t)(1-t^2)^{-1/2}\,\mathrm{d}t$$

$$M_n^{22}(x) = \frac{1}{\pi}\int_{-1}^{1} K_{22}(x,t)C_{2n-1}(t)(1-t^2)^{-1/2}\,\mathrm{d}t$$

where $C_n(t)$ are the Gegenbauer polynomials which for this special case can be reduced to the Chebyshev polynomials of the first (T_n) and the second (U_n) kind. Hence, the following formulae are valid:

$$\frac{1}{\pi}\int_{-1}^{1} C_n(t)(1-t^2)^{-1/2}\frac{\mathrm{d}t}{t-x} = C_{n-1}(x), \quad |x|<1 \tag{3.7.10}$$

Tables 3.7.1-3.7.4 show the numerical results for $p = 1$, $a = 1$ and $d/a = 0.4$, $d/a = 0.8$, $d/a=1.2$, $d/a = 2.0$, respectively.

Table 3.7.1. Stress intensity factors for *d*/*a* = 0.4

N		
2	2.54368	0.76485
3	2.56923	0.75062
4	2.59487	0.73638
5	2.59523	0.73687
6	2.59565	0.73757
7	2.59562	0.73754
8	2.59559	0.73752
9	2.59560	0.73750

Table 3.7.2. Stress intensity factors for *d*/*a* = 0.8

N		
2	1.63856	0.26779
3	1.64977	0.26956
4	1.66102	0.27154
5	1.66096	0.27153
6	1.66091	0.27152
7	1.66090	0.27150

Table 3.7.3. Stress intensity factors for *d*/*a* = 1.2

N		
2	1.36604	0.12230
3	1.36914	0.12283
4	1.37225	0.12338
5	1.37223	0.12339
6	1.37220	0.12340
7	1.37220	0.12340

Table 3.7.4. Stress intensity factors for *d*/*a* = 2.0

N		
2	1.16148	0.03660
3	1.16177	0.03665
4	1.16210	0.03668
5	1.16209	0.03667
6	1.16209	0.03667

3.8. Stress Intensity Factors near Straight Cracks in a Bimaterial Infinite and Isotropic Solid under Antiplane Shear

The finite-part singular integrals will be applied to other fracture mechanics problems, by determining the stress intensity factors in the neighbourhood of a straight crack in a bimaterial infinite and isotropic solid, under antiplane shear (Figure 3.8.1). The above problem has been previously solved by F. Erdogan and T. S. Cook [31] and J. L. Bassani and F. Erdogan [32] by using some numerical methods, while in the present section a closed-form solution is investigated.

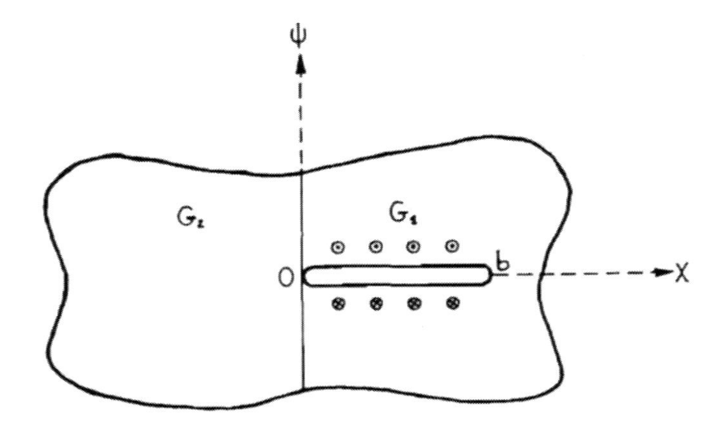

Figure 3.8.1. A straight crack in a bimaterial infinite and isotropic solid, under antiplane shear.

This problem is reduced to the solution of the following singular integral equation [10], [31]:

$$\frac{1}{\pi}\int_{0}^{b}\left[\frac{1}{t-x}+\frac{\zeta}{t+x}\right]g(t)\,\mathrm{d}t=\frac{2}{G_1}q,\quad 0<x<b \tag{3.8.1}$$

with:

$$\zeta=\frac{G_1-G_2}{G_1+G_2} \tag{3.8.2}$$

in which q is the shear loading and G_1 and G_2 are respectively the shear modulus of each of the two bounded half planes of the bimaterial solid.

Thus, by integrating (3.8.1) by parts and using the appropriate boundary conditions, one has:

$$\frac{1}{\pi}\oint_{0}^{b}\left[\frac{1}{(t-x)^2}+\frac{\zeta}{(t+x)^2}\right]h(t)\,\mathrm{d}t=\frac{2}{G_1}q,\quad 0<x<b \tag{3.8.3}$$

with $g(t)=h'(t)$, where the integral in (3.8.3) is to be interpreted in the finite-part sense.

Furthermore, the unknown function $h(t)$ may be written as the following:

$$h(t)=w(t)\delta(t) \tag{3.8.4}$$

where $w(t)$ is the weight function defined as:

$$w(t) = t^{\beta}(b-t)^{1/2} \tag{3.8.5}$$

and β is the first root of the relation:

$$\cos(\pi\beta) = \zeta \tag{3.8.6}$$

By applying further the method given by Bückner [1], we obtain the following closed form solution of (3.8.1):

$$g(s) = \frac{q}{G_1 \sin(\pi\varepsilon/2)} \left[\begin{array}{l} \left(\dfrac{s}{1+(1-s^2)^{1/2}}\right)^{\varepsilon} \left(\dfrac{\varepsilon}{(1-s^2)^{1/2}} + 1\right) + \\ + \left(\dfrac{s}{1+(1-s^2)^{1/2}}\right)^{-\varepsilon} \left(\dfrac{\varepsilon}{(1-s^2)^{1/2}} - 1\right) \end{array} \right] \tag{3.8.7}$$

with:

$$\varepsilon = 1-\beta, \quad s = t/b, \quad 0 < t < b \tag{3.8.8}$$

The exact expressions for the stress intensity factors *k(0)* and *k(b)* are given by the following relations:

$$k(b) = -\frac{G_1}{2} \lim_{t \to b} [2(b-t)]^{1/2} g(t) \tag{3.8.9}$$

$$k(0) = -\left(\frac{G_1 G_2}{2}\right)^{1/2} \lim_{t \to 0} t^{1-\beta} g(t) \tag{3.8.10}$$

Hence, by inserting (3.8.7), in (3.8.9) and (3.8.10), one has:

$$k(b) = -q\sqrt{b}\,\frac{1-\beta}{\cos(\pi\beta/2)} \tag{3.8.11}$$

$$k(0) = \frac{q}{2^{\beta}} \left(\frac{2G_2}{G_1}\right)^{1/2} \frac{\beta b^{1-\beta}}{\cos(\pi\beta/2)} \tag{3.8.12}$$

Thus, the stress intensity factors $k(0)$ and $k(b)$ may be written, in the following reduced forms:

$$k^*(b) = \frac{\sqrt{2}k(b)}{q\sqrt{b}} \tag{3.8.13}$$

$$k^*(0) = \frac{2^{1-\beta} k(0)}{q b^{1-\beta}} \qquad (3.8.14)$$

Table 3.8.1 shows the values of the stress intensity factors $k^*(b)$ and $k^*(0)$ for many solids like glass-iron, iron-glass, cast iron-steel, steel-cast iron, aluminium-copper, copper-aluminium, epoxy-glass, glass-epoxy, aluminium-cast iron, cast iron-aluminium, copper-steel and steel-copper.

Table 3.8.1. Stress intensity factors for the straight crack of Figure 3.8.1

		ζ	β
Solids	G_2/G_1	Eq. (3.8.2)	Eq. (3.8.6)
Glass-Iron	3.2520	-0.52964	0.67767
Iron-Glass	0.3075	0.52964	0.32233
Cast Iron-Steel	2.1053	-0.35593	0.61584
Steel-Cast Iron	0.4750	0.35596	0.38416
Aluminium-Copper	1.4444	-0.18182	0.55820
Copper-Aluminium	0.6923	0.18182	0.44180
Epoxy-Glass	20.5000	-0.90698	0.86162
Glass-Epoxy	0.0488	0.90698	0.13838
Aluminium-Cast Iron	1.4074	-0.16923	0.55413
Cast Iron-Aluminium	0.7105	0.16923	0.44587
Copper-Steel	2.0513	-0.34454	0.61197
Steel-Copper	0.4875	0.34454	0.38803

	$k^*(b)$	$k^*(0)$
Solids	Eq. (3.8.13)	Eq. (3.8.14)
Glass-Iron	0.939965	2.785715
Iron-Glass	1.095859	0.369767
Cast Iron-Steel	0.957370	1.896487
Steel-Cast Iron	1.057736	0.533963
Aluminium-Copper	0.976860	1.368360
Copper-Aluminium	1.026936	0.733105
Epoxy-Glass	0.907445	15.496118
Glass-Epoxy	1.247879	0.073090
Aluminium-Cast Iron	0.978362	1.338203
Cast Iron-Aluminium	1.024921	0.749305
Copper-Steel	0.958577	1.853932
Steel-Copper	1.055531	0.545765

Table 3.8.2. Stress intensity factors for the straight crack of Figure 3.8.1, for Epoxy-Aluminium

	$k^*(0)$
Numerical ~ Ref. [29]	13.13
Numerical ~ Ref. [30]	14.00
Exact (eq. (3.8.14))	17.368088

Beyond the above, Table 3.8.2 shows the exact values of the stress intensity factor $k^*(0)$ for epoxy-aluminium ($G_2/G_1 = 23.077$, $\zeta = -0.91694$, $\beta = 0.86935$), which are compared with the numerical results obtained by Erdogan and Cook [31] and Bassani and Erdogan [32].

3.9. Fracture Mechanics of Straight Cracks in an Infinite and Isotropic Medium

Let us consider a straight crack of length $2b$ inside an infinite and isotropic solid (see : Figure 3.9.1)

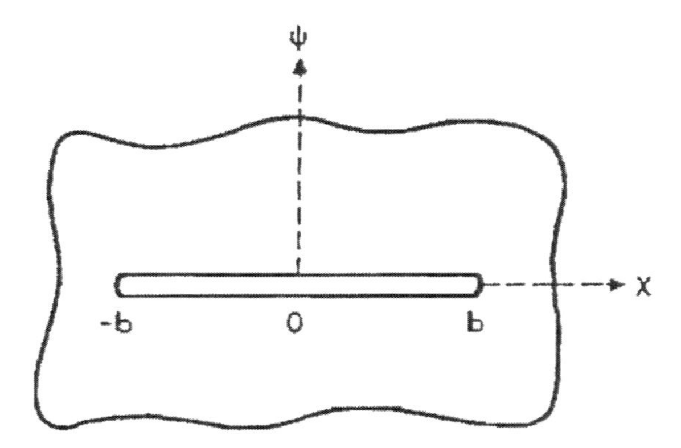

Figure 3.9.1. A straight crack inside an infinite & isotropic solid.

For the determination of the stress field near the crack edges, then the following singular integral equation has to be solved [8], [13]:

$$\frac{1}{\pi}\int_{-1}^{1}\frac{g(t)\,\mathrm{d}t}{t-x} = p(x), \quad x \in (-1,1) \tag{3.9.1}$$

in which the unknown edge-dislocation along the crack $g(t)$, must satisfy the condition of continuity:

$$\int_{-1}^{1} g(t)\,\mathrm{d}t = 0 \tag{3.9.2}$$

and $p(x)$ is the loading distribution along the crack edges.

By following Muskhelishvili [27], the Cauchy-type singular integral will be used:

$$G(z) = \frac{1}{2\pi i}\int_{-1}^{1}\frac{g(t)\,\mathrm{d}t}{t-z}, \quad z = x + iy \tag{3.9.3}$$

By using further the Plemelj formulas [27] one has:

$$G^{+}(x)-G^{-}(x)=g(x)$$
$$G^{+}(x)+G^{-}(x)=\frac{1}{\pi i}\int_{-1}^{1}\frac{g(t)\,\mathrm{d}t}{t-x} \tag{3.9.4}$$

Hence, the problem reduces to the determination of the function $G(z)$ which satisfies the relation:

$$G^{+}(x)+G^{-}(x)=-ip(x) \tag{3.9.5}$$

The solution of (3.9.5) is:

$$G(z)=\frac{1}{2\pi i(z^{2}-1)^{1/2}}\int_{-1}^{1}\frac{(1-t^{2})^{1/2}g(t)\,\mathrm{d}t}{t-z}-\frac{C}{2i(z^{2}-1)^{1/2}} \tag{3.9.6}$$

where C is a real constant.

By using (3.9.6), then the solution of (3.9.1) is obtained:

$$g(x)=-\frac{1}{\pi(1-x^{2})^{1/2}}\int_{-1}^{1}\frac{(1-t^{2})^{1/2}p(t)\,\mathrm{d}t}{t-x}+\frac{C}{(1-x^{2})^{1/2}} \tag{3.9.7}$$

The corresponding solution which is bounded and zero at the ends is:

$$G(z)=-\frac{1}{2\pi i}(z^{2}-1)^{1/2}\int_{-1}^{1}\frac{g(t)\,\mathrm{d}t}{(1-t^{2})^{1/2}(t-z)} \tag{3.9.8}$$

and:

$$g(x)=-\frac{(1-x^{2})^{1/2}}{\pi}\int_{-1}^{1}\frac{p(t)\,\mathrm{d}t}{(1-t^{2})^{1/2}(t-x)} \tag{3.9.9}$$

in which the loading distribution along the crack edges $p(t)$ must satisfy the condition:

$$\int_{-1}^{1}\frac{p(t)\,\mathrm{d}t}{(1-t^{2})^{1/2}}=0 \tag{3.9.10}$$

We assume further that $p(x)$ is discontinuous at the point $x = 0$ along the straight crack. Hence, a logarithmic singularity of $g(t)$ appears at $t = 0$ and it can be replaced by the relation:

$$g(t) = w(t) \cdot l(t) \tag{3.9.11}$$

where by $w(t)$ is denoted the logarithmic weight function:

$$w(t) = \ln(1/t) \tag{3.9.12}$$

Beyond the above, by using (3.9.11) and (3.9.12), the singular integral equation (3.9.1) becomes:

$$\frac{2x}{\pi} \int_0^1 l(t) \ln(1/t) \frac{\mathrm{d}t}{t^2 - x^2} = p(x), \quad x \in [0,1) \tag{3.9.13}$$

and the condition of continuity:

$$\int_0^1 l(t) \ln(1/t) \, \mathrm{d}t = 0 \tag{3.9.14}$$

As it is easily seen a closed form solution of eq. (3.9.7) is not simple, but it can be solved by using the numerical methods described in the previous chapter.

On the other hand, in order to make a comparison between the theoretical values and the numerical results obtained by the new method, we consider the simple case where the load is given by the simple formula $p(x) = \mathrm{sign}(x)$. For this case, the determination of a closed form solution of eq. (3.9.7) is possible. Thus, (3.9.7) reduces to:

$$l(x) = -\frac{2}{\pi} \tan h^{-1} (1 - x^2)^{1/2} + B(1 - x^2)^{-1/2} \tag{3.9.15}$$

in which B is a real constant, and finally by using (3.9.12) one obtains:

$$l(0) = -2/\pi = -0.636619772 \tag{3.9.16}$$

Moreover, by using Radau's numerical method described in the last section, we obtain the numerical results shown in Table 3.9.1. Finally, as it is seen from Table 3.9.1 the numerical results converge to the theoretical value (3.9.16) with very good accuracy.

Table 3.9.1. Numerical results for the intensity of the logarithmic singularity at $t = 0$ in the crack of Figure 3.9.1

n	I(0)
4	-0.5822
6	-0.5996
8	-0.6021
10	-0.6055
12	-0.6073
14	-0.6095
16	-0.6100
20	-0.6132

3.10. Stress Intensity Factors in the Edge of a Periodic Array of Cracks along a Straight Line in an Infinite and Isotropic Solid

As an application of the previous theory, the stress intensity factors will be determined in the edge of a periodic array of cracks of length $2a$, along a straight line with period equal to d, in an infinite and isotropic solid (Figure 3.10.1). The same problem has been previously solved by A. P. Datsyshin and M. P. Savruck [26], while here will be solved by a singular integral equations method [7].

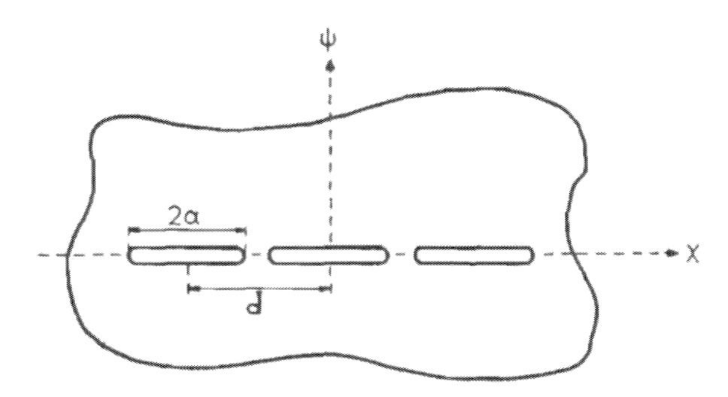

Figure 3.10.1. A periodic array of cracks of length $2a$, along a straight line with period equal to d, in an infinite and isotropic solid.

The above problem reduces to the solution of the following singular integral equation of the first kind [7]:

$$\frac{1}{\pi}\int_{-1}^{1}\frac{\varphi(t)}{t-x}\,\mathrm{d}t+\int_{-1}^{1}K_1(t,x)\varphi(t)\,\mathrm{d}t=f(x) \qquad (3.10.1)$$

where the kernel $K_1(x,t)$ is given by the following formula:

$$K_1(x,t)=\frac{a}{d}\cot\left[\frac{\pi a(t-x)}{d}\right]-\frac{1}{\pi(t-x)} \tag{3.10.2}$$

and the pressure distribution $f(x)$ along the crack faces is equal to:

$$f(x)=px^6 \tag{3.10.3}$$

in which p is a normal loading intensity.

In order the singular integral equation (3.10.1) to be solved the Gauss-Chebyshev and the Lobatto-Chebyshev integration rule was used, which was described in Chapter 2. Hence, the stress intensity factor $k/pa^{1/2}$ is directly evaluated at the crack tips.

Table 3.10.1 shows the values of the stress intensity factors $k/pa^{1/2}$ at the crack tips, by using the Gauss-Chebyshev integration rule and the Lobatto-Chebyshev rule. As it can be seen from the above table the results of the two methods compare very well.

Table 3.10.1. Stress intensity factors $k/pa^{1/2}$ at the crack tips of the periodic array of cracks of Fig. 3.10.1

	$a/d = 0.1$		$a/d = 0.4$	
n	Gauss-Chebyshev	Lobatto-Chebyshev	Gauss-Chebyshev	Lobatto-Chebyshev
2	0.299738	0.283646	0.368064	0.371848
3	0.313831	0.313831	0.362172	0.362605
4	0.313831	0.313831	0.361730	0.361733

3.11. Stress Intensity Factors in the Edge of a Periodic Array of Parallel Straight Cracks in an Infinite and Isotropic Solid

The fracture mechanics theory will be further applied to the determination of the stress intensity factors on the edge of a periodic array of parallel straight cracks of length $2a$ in an infinite and isotropic solid, in which the distance between two successive cracks is denoted by d (Figure 3.11.1).

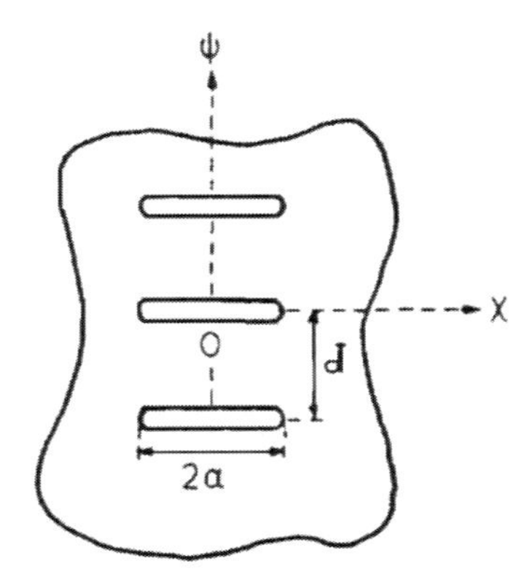

Figure 3.11.1. A periodic array of parallel straight cracks of length $2a$ in an infinite and isotropic solid.

The above elasticity problem reduces to the solution of the following singular integral equation of the first kind [7]:

$$\frac{1}{\pi}\int_{-1}^{1}\frac{\varphi(t)}{t-x}\mathrm{d}t+\int_{-1}^{1}K_{II}(t,x)\varphi(t)\,\mathrm{d}t=f(x) \tag{3.11.1}$$

where the kernel $K_{II}(x,t)$ is valid as:

$$K_{II}(x,t)=\frac{a}{d}\left[\frac{2\cot h\left(\dfrac{\pi a(t-x)}{d}\right)-\dfrac{\pi a(t-x)}{d}\operatorname{cosec}^2}{\left\{\dfrac{\pi a(t-x)}{d}\right\}}\right]-\frac{1}{\pi(t-x)} \tag{3.11.2}$$

and $f(x)$ denotes the pressure distribution along the crack faces given by (3.10.3).

The Gauss-Chebyshev integration rule was used for the numerical solution of the singular integral equation (3.11.1) and hence the stress intensity factor $k/pa^{1/2}$ is directly evaluated at the crack tips.

Table 3.11.1 shows the values of the stress intensity factors $k/pa^{1/2}$ at the crack tips, by using the Gauss-Chebyshev integration rule and the Lobatto-Chebyshev rule. As it can be seen in this table, the results of the two methods compare very well.

Table 3.11.1. Stress intensity factors $k/pa^{1/2}$ at the crack tips of the periodic array of parallel straight cracks of Figure 3.11.1

	$a/d = 0.1$		$a/d = 0.4$	
n	Gauss-Chebyshev	Lobatto-Chebyshev	Gauss-Chebyshev	Lobatto-Chebyshev
2	0.291805	0.274823	0.258563	0.232154
3	0.308928	0.308929	0.285178	0.285221
4	0.308927	0.308928	0.285137	0.285137

REFERENCES

[1] H.F. Bückner, On a class of singular integral equations, *J. Math. Anal. Appl.* 14, 392-426 (1966).

[2] H.F. Bückner, Field singularities and related integral representations, in *Methods of Analysis and Solutions of Crack Problems,* (Edited by G.C. Sih), Vol.1, p.p. 239-314, Noordhoff, Leyden, The Netherlands (1973).

[3] F. Erdogan, Approximate solutions of systems of singular integral equations, *SIAM J. Appl. Math.* 17, 1041-1059 (1969).

[4] A.C. Kaya and F. Erdogan, On the solution of integral equations with strongly singular kernels, *Q. Appl. Math.* 45, 105-122 (1987).

[5] A.C. Kaya and F. Erdogan, On the solution of integral equations with a generalized Cauchy kernel, *Q. Appl. Math.* 45, 455-469 (1987).

[6] E.G. Ladopoulos, On the numerical solution of the finite-part singular integral equations of the first and the second kind used in fracture mechanics, *Comp. Meth. Appl. Mech. Engng* 65, 253-266 (1987).

[7] E.G. Ladopoulos, On the numerical evaluation of the general type of finite-part singular integrals and integral equations used in fracture mechanics, *J. Engng Fract. Mech.* 31, 315-337 (1988).

[8] E.G. Ladopoulos, The general type of finite-part singular integrals and integral equations with logarithmic singularities used in fracture mechanics, *Acta Mech.* 75, 275-285 (1988).

[9] E.G. Ladopoulos, Systems of finite-part singular integral equations in Lp applied to crack problems, *J. Engng Fract. Mech.* 48, 257-266 (1994).

[10] E.G. Ladopoulos, New aspects for the generalization of the Sokhotski-Plemelj formulae for the solution of finite-part singular integrals used in fracture mechanics, *Int. J. Fract.* 54, 317-328 (1992).

[11] E.G. Ladopoulos, V.A.Zisis and D. Kravvaritis, Singular integral equations in Hilbert space applied to crack problems, *Theor. Appl. Fract. Mech.* 9, 271-281 (1988).

[12] E.G. Ladopoulos, D. Kravvaritis and V.A. Zisis, Finite-part singular integral representation analysis in Lp of two-dimensional elasticity problems, *J. Engng Fract. Mech.* 43, 445-454 (1992).

[13] F. Erdogan, G.D. Gupta and T.S. Cook, Numerical solution of singular integral equations, in *Methods of Analysis and Solutions of Crack Problems* (Edited by G.C.Sih), Vol. 1, pp. 368-425, Noordhoff, Leyden, The Netherlands (1973).

[14] G.M.L. Gladwell and A.H. England, Orthogonal polynomial solutions to some mixed boundary-value problems in elasticity theory, *Q. J. Mech. Appl. Math.* 30, 175-185 (1977).

[15] R.P. Gilbert and R. Magnanini, The boundary integral method for two-dimensional orthotropic materials, *J. Elasticity* 18, 61-82 (1987).

[16] D.A. Hills, D.N. Dai, P.A. Kelly and A.M. Korsunsky, *Singular Integral Equations in the Mechanics of Fracture*, Kluwer, Dordrecht (1995).

[17] A.M. Korsunsky, Gauss-Chebyshev quadrature formulae for strongly singular integrals, *Q. Appl. Math.* 56, 461-472 (1998).

[18] E.G. Ladopoulos, On a new integration rule with the Gegenbauer polynomials for singular integral equations, used in the theory of elasticity, *Ing. Archiv* 58, 35-46 (1988).

[19] V.A. Zisis and E.G. Ladopoulos, Singular integral approximations in Hilbert spaces for elastic stress analysis in a circular ring with curvilinear cracks, *Indus. Math.* 39, 113-134 (1989).

[20] V.V. Panasyuk, M.P. Savruk and A.P. Datsyshin, *Stress Distribution near Cracks in Plates and Shells,* Naukova Dumka, Kiev (1976).

[21] A.N. Teong and D.L.Clements, A boundary integral equation method for the solution of a class of crack problems, *J. Elasticity* 17, 9-21 (1987).

[22] U. Zastrow, Solution of the anisotropic elastostatical boundary value problems by singular integral equations, *Acta Mech.* 44, 59-71 (1982).
[23] U. Zastrow, Numerical plane stress analysis by integral equations based on the singularity method, *Solid Mech. Arch.* 10, 113-128 (1985).
[24] I.N. Sneddon and S.C. Das, The problem of the unsymmetrical cruciform crack, in *Trends in Elasticity and Thermoelasticity,* Noordhoff, Groningen, The Netherlands (1971).
[25] D.P. Rooke and I.N. Sneddon, The crack energy and the stress intensity factor for a cruciform crack deformed by internal pressure, *Int. J. Engng Sci.* 7, 1079-1089 (1969).
[26] A.P. Datsyshin and M.P. Savruk, A system of arbitrarily orientated cracks in elastic solids, *J. Appl. Math. Mech.* 37, 306-313 (1973).
[27] N.I. Muskhelishvili, *Some Basic Problems of the Mathematical Theory of Elasticity,* Noordhoff, Groningen, The Netherlands (1953).
[28] S.G. Lekhnitskii, *Theory of Elasticity of an Anisotropic Elastic Body,* Holden - Day, San Francisco (1963).
[29] S.G. Lekhnitskii, *Anisotropic Plates,* Gordon and Breach, New York (1968).
[30] F. Erdogan and G.D. Gupta, The stress analysis of multi-layered composites with a flow, *Int. J. Sol. Struct.* 7, 39-61 (1971).
[31] F. Erdogan and T.S. Cook, Antiplane shear crack terminating at and going through a bimaterial interface, *Int. J. Fract.* 10, 227-240 (1974).
[32] J.L. Bassani and F. Erdogan, Stress intensity factors in bonded half plane containing inclined cracks and subjected to antiplane shear loading, *Int. J. Fract.* 15, 145-158 (1979).

Chapter 4

AERODYNAMICS BY SINGULAR INTEGRAL EQUATIONS

4.1. INTRODUCTION

Another major field of engineering mechanics and mathematical physics in which the finite-part singular integral equations are applied, is fluid dynamics and aerodynamics. Such systems of singular integral equations are evaluated only by computational recipes because of the big complication of their form, as closed form solutions are available only in very seldom cases.

A very effective method for solving numerically this type of singular integral equations is the direct method which consists in reducing such an equation (or systems of equations) to a system of linear algebraic equations, by using an appropriate numerical integration rule on a properly selected set of collocation points.

Several studies of the generalized two-dimensional aerodynamics began very early, in the 1930. Theories for general aerodynamic problems have been obtained by V.V. Golubev [1], T. von Karman and J.M. Burgers [2], H. Schmidt [3] and K. Schröder [4], [5].

In the 1940 the two-dimensional aerodynamic problems were advanced by the work of J. Weissinger [6], H. Küssner and L. Schwarz [7], L.G. Magnaradze [9], [10], I.N. Vekua [11], [12] and H. Schöngen [13].

Recently, some studies have been published on the application of the singular integral equations in aerodynamics. Among them we shall mention the following authors : S.R. Bland et al. [14], [15], J. Blackwell and G. Pounds [16], J.A. Fromme and M.A. Golberg [17] - [22], M.A. Golberg, M. Lea and G. Miel [23], M.A. Golberg [24], E.G. Ladopoulos [25], W.F. Moss [26], [27], D.J. Salmond [28], M.H. Williams [29], E. Kraft and C. Lo [30], M. Mokry [31] and E. Nissim and I. Lottati [32].

By the present chapter the generalized airfoil equation is investigated and solved, which presents the pressure acting on a planar airfoil undergoing simple amplitude oscillations about the central plane of a two-dimensional ventilated wind tunnel.

4.2. Two-Dimensional Aerodynamics Applications of Planar Airfoils

Let a planar airfoil undergoing simple amplitude oscillations about a center plane of a two-dimensional ventilated wind tunnel (Figure 4.2.1). Then, by removing the walls to infinity, a very important special case exists which gives free air conditions.

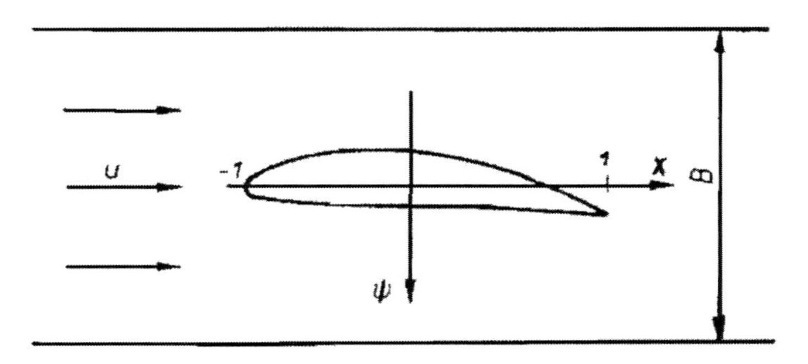

Figure 4.2.1. A planar airfoil in a two-dimensional ventilated wind tunnel.

Beyond the above the flow is assumed to be inviscid and strictly subsonic and thus the following unsteady wave equation is valid: [19], [25]

$$\nabla^2 \xi - M^2 \left(\frac{\vartheta}{\vartheta x} + ik \right)^2 \xi = 0 \tag{4.2.1}$$

in which ξ denotes the perturbation velocity potential, M the freestream Mach number and k the reduced frequency:

$$k = \frac{\omega d}{u} \tag{4.2.2}$$

where ω is the frequency of the simple harmonic motion of the airfoil, d its semi-chord and u the free stream velocity.

Moreover, the nondimensional pertubation pressure p is given by the following relation:

$$p = -2\left(\frac{\vartheta}{\vartheta x} + ik \right)\xi \tag{4.2.3}$$

with the boundary conditions:

$$p(x,0) = \begin{cases} 0, & |x| \geq 1 \\ -1/2\Delta p(x), & |x| < 1 \end{cases} \tag{4.2.4}$$

in which Δp denotes the lifting pressure jump across the airfoil. The relation between the downwash velocity w and the pressure potential ξ is equal to:

$$w(x) = \left.\frac{\vartheta\xi}{\vartheta y}\right|_{y=0}, \quad |x| < 1 \tag{4.2.5}$$

Hence, the downwash velocity w is related to the potential ξ as follows: [14]

$$w(x, y, t) = \frac{1}{u}\int_{-\infty}^{x} \xi_y\left(\mu, y, t - d\,\frac{x-\mu}{u}\right) \mathrm{d}\,\mu \tag{4.2.6}$$

where t denotes the time and μ the ventilation coefficient.

By using the Fourier transforms:

$$\begin{aligned} \Xi(s, y) &= \int_{-\infty}^{\infty} e^{-ixt}\,\xi(x, y)\,\mathrm{d}\,x \\ \xi(x, y) &= \int_{-\infty}^{\infty} e^{ixs}\,\Xi(s, y)\,\mathrm{d}\,s \end{aligned} \tag{4.2.7}$$

the pressure potential will be given by the following relation:

$$\xi(x, y) = \frac{1}{4\pi\rho_0}\int_{-\infty}^{\infty} e^{ixs} f(s) \int_{-1}^{1} e^{-is\zeta}\,\Delta p(\zeta)\,\mathrm{d}\,\zeta\,\mathrm{d}\,s \tag{4.2.8}$$

where:

$$f(s) = \frac{\sin ha(B/2 - y) + ca\cos ha(B/2 - y)}{\sin h(a\,B/2) + ca\cos h(a\,B/2)} \tag{4.2.9}$$

in which c denotes the porosity coefficient, B the tunnel height and ρ_0 the free stream density.

Furthermore, in eq. (4.2.9) the parameter a is valid as:

$$a(s) = (\beta^2 s^2 - 2M^2 gs - M^2 g^2)^{1/2} \tag{4.2.10}$$

where g is the complex reduced frequency and $\beta = \sqrt{1 - M^2}$.

By combining therefore eqs (4.2.6) and (4.2.8), one has:

$$\frac{w(x, y)}{u} = \frac{1}{4\pi\rho_0 u^2}\int_{-\infty}^{x} e^{-ig(x-\mu)}\,\frac{\vartheta}{\vartheta y}\int_{-\infty}^{\infty} e^{i\mu t} f(s) \int_{-1}^{1} e^{-is\zeta}\,\Delta p(\zeta)\,\mathrm{d}\,\zeta\,\mathrm{d}\,s\,\mathrm{d}\,\mu \tag{4.2.11}$$

By taking the derivative and interchanging the orders of integration, we obtain:

$$w(x,y) = \frac{2}{\rho_0 u} \int_{-1}^{1} \Delta p(\zeta) K(M, g, x-\zeta, y, B, c) \, \mathrm{d}\zeta \tag{4.2.12}$$

in which the kernel function K is given by the formula:

$$K = -\frac{1}{8\pi} \int_{-\infty}^{\infty} a e^{-ig(x-\zeta)} \frac{\cos ha(B/2-y) + ca \sin ha(B/2-y)}{\sin h(aB/2) + ca \cos h(aB/2)} \times \\ \times \int_{-\infty}^{x-\zeta} e^{i(s+g)\mu} \, \mathrm{d}\mu \, \mathrm{d}s \tag{4.2.13}$$

For steady ($g = 0$), incompressible ($M = 0$) flow and in free air (no tunnel walls $B = \infty$), the kernel takes the simple form:

$$K(x) = 1/x \tag{4.2.14}$$

For this case, with $y = 0$, from eq. (4.2.12) results the following singular integral equation:

$$w(x) = \frac{1}{2\pi\rho_0} \int_{-1}^{1} \frac{\Delta p(\zeta)}{\zeta - x} \mathrm{d}\zeta \tag{4.2.15}$$

By using further the Kutta boundary condition of a smooth flow at the airfoil trailing edge:

$$\lim_{x \to 1} \frac{2\Delta p(x,t)}{\rho_0 u^2} = 0 \tag{4.2.16}$$

then eq. (4.2.15) has the following closed form solution:

$$\Delta p(\zeta) = -\frac{2\rho_0 u}{\pi} \left(\frac{1-\zeta}{1+\zeta} \right)^{1/2} \int_{-1}^{1} \frac{w^*(x) w(x)}{x - \zeta} \mathrm{d}x \tag{4.2.17}$$

with the weight function $w^*(x) = (1+x)^{1/2}(1-x)^{-1/2}$.

Hence, by putting the pressure factor:

$$p(\zeta) = -\frac{4}{\pi}\int_{-1}^{1} w^*(x)\frac{w(x)}{u}\frac{\mathrm{d}x}{x-\zeta} \tag{4.2.18}$$

then eq.. (4.2.17) can be written as follows:

$$\Delta p(\zeta) = \frac{1}{2}\rho_0 u^2\left(\frac{1-\zeta}{1+\zeta}\right)^{1/2} p(\zeta) \tag{4.2.19}$$

The pressure factor $p(\zeta)$ in eq. (4.2.18) is continuous on [-1,1] if $w(x)/u$ is also continuous.

Numerical Evaluation of the Airfoil Equation

For the numerical evaluation of the airfoil equation (4.2.18) the Gauss-Chebyshev numerical integration rule will be used, while solving the same problem, S. R. Bland [14] has used the Gauss-Jacobi rule.

Let the following singular integral:

$$\Phi(\zeta) = \frac{1}{2\pi i}\int_{-1}^{1}\frac{w^*(x)\varphi(x)}{x-\zeta}\mathrm{d}x \tag{4.2.20}$$

in which $w^*(x)$ is the weight function defined in the interval [-1,1], $\varphi(x)$ is an analytic function without poles in a domain Ω containing the interval [-1,1] and $\Phi(\zeta)$ is a sectionally analytic function in the whole complex plane except [-1,1].

In order to solve numerically the singular integral (4.2.20), we consider the following contour integral on a curve C surrounding the interval [-1,1]:

$$\Phi_0 = \frac{1}{2\pi i}\int_C \frac{\varphi(\zeta')}{(\zeta'-x)(\zeta'-\zeta)m_n(\zeta')}\mathrm{d}\zeta' \tag{4.2.21}$$

where:

$$m_n(\zeta) = \prod_{k=1}^{n}(\zeta - x_k) \tag{4.2.22}$$

in which x_k are the abscissae.

Hence, by applying the Cauchy residue theorem to the integral (4.2.20), we obtain:

$$2\pi i \Phi(\zeta) = \int_{-1}^{1} \frac{w^*(x)\varphi(x)}{x-\zeta} \mathrm{d}\, x = \\ = \sum_{k=1}^{n} A_k \frac{\varphi(x_k)}{x_k - \zeta} - 2\varphi(\zeta)\frac{d_n(\zeta)}{m_n(\zeta)} + E_n \tag{4.2.23}$$

where the error function E_n is equal to:

$$E_n = \frac{1}{\pi i} \int_C \frac{\varphi(\zeta')}{\zeta' - \zeta} \frac{d_n(\zeta')}{m_n(\zeta')} \mathrm{d}\zeta' \tag{4.2.24}$$

A_k are the weights and $d_n(\zeta)$ is given by the relation:

$$d_n(\zeta) = -\frac{1}{2} \int_{-1}^{1} w^*(x) \frac{m_n(x)}{x-\zeta} \mathrm{d}\, x \tag{4.2.25}$$

By using further the Gauss-Chebyshev numerical integration rule with the weight function $w^*(x) = (1+x)^{\pm 1/2}(1-x)^{\pm 1/2}$, then the relation (4.2.23) can be written as:

$$\int_{-1}^{1} \frac{w^*(x)\varphi(x)}{x-\zeta} \mathrm{d}\, x = \sum_{k=1}^{n} A_k \frac{\varphi(x_k)}{x_k - \zeta} - 2\varphi(\zeta) R_n(\zeta) + E_n \tag{4.2.26}$$

for $\zeta \neq x_m$, $m = 1, 2,\ldots, n$, and:

$$\int_{-1}^{1} \frac{w^*(x)\varphi(x)}{x-\zeta} \mathrm{d}\, x = \sum_{\substack{k=1 \\ k\neq m}}^{n} A_k \frac{\varphi(x_k)}{x_k - \zeta} + A_m \varphi'(\zeta) - 2\varphi(\zeta) G_n(\zeta) + E_n \tag{4.2.27}$$

for $\zeta = x_m$, $m = 1, 2,\ldots, n$, where:

$$R_n(\zeta) = -\frac{\pi U_{n-1}(\zeta)}{n T_n(\zeta)}, \quad \zeta \neq x_m, \quad m = 1,2,\ldots,n \tag{4.2.28}$$

and:

$$G_n(\zeta) = -\frac{\pi}{2} \frac{U_{n-2}(\zeta)}{T_{n-1}(\zeta)} + \frac{2n-1}{4} A_m \frac{\zeta}{1-\zeta^2}, \quad \zeta = x_m, \quad m = 1,2,\ldots,n \tag{4.2.29}$$

in which $T_n(\zeta)$ and $U_n(\zeta)$ denote the Chebyshev polynomials of the first and the second kind and degree n, respectively, expressible in terms of trigonometric functions as follows:

$$T_n(\zeta) = \cos n\vartheta$$

$$U_{n-1}(\zeta) = \frac{\sin n\vartheta}{\sin\vartheta} \tag{4.2.30}$$

$$\zeta = \cos\vartheta$$

In eqs. (4.2.26) and (4.2.27) ζ is not permitted to coincide with the endpoints -1 or 1 of the integration interval.

As an application of the airfoil equation (4.2.18), we consider the case where the downwash is valid as:

$$\frac{w(x)}{u} = \begin{cases} 0, & x \leq 0 \\ x, & x > 0 \end{cases} \tag{4.2.31}$$

Therefore, by using the Gauss-Chebyshev numerical integration rule given by eqs. (4.2.26) and (4.2.27), it is possible to compute the airfoil equation (4.2.18). The same equation was computed by S. R. Bland [14], while using the Gauss-Jacobi rule.

Figure 4.2.2 shows the pressure distribution $p(\zeta)$ for downwash given by eq. (4.2.31).

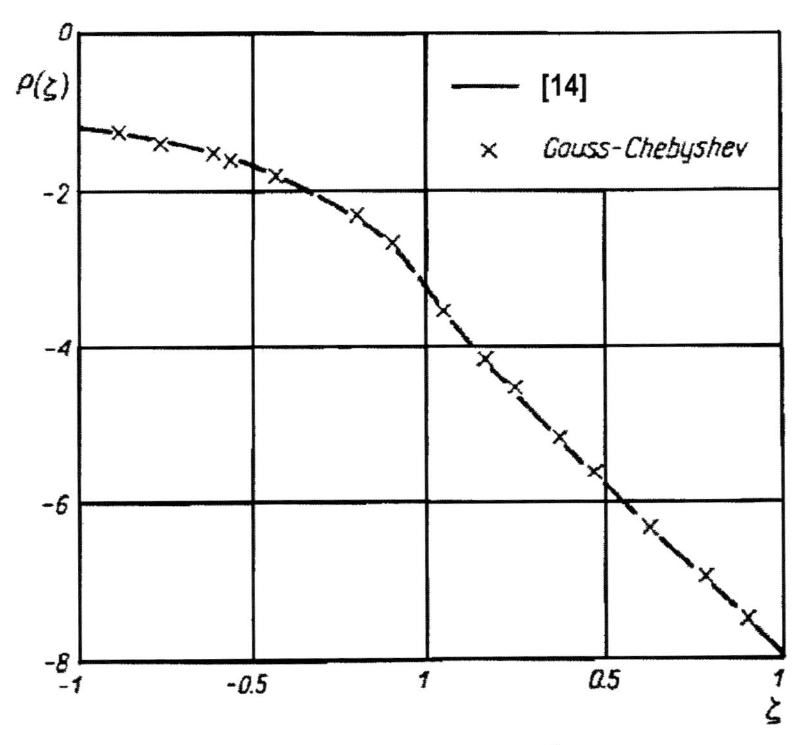

Figure 4.2.2. Pressure distribution $p(\zeta)$ for downwash $w(x)/u = \begin{cases} 0, x \leq 0 \\ x, x > 0 \end{cases}$ for the planar airfoil of Figure 4.2.1.

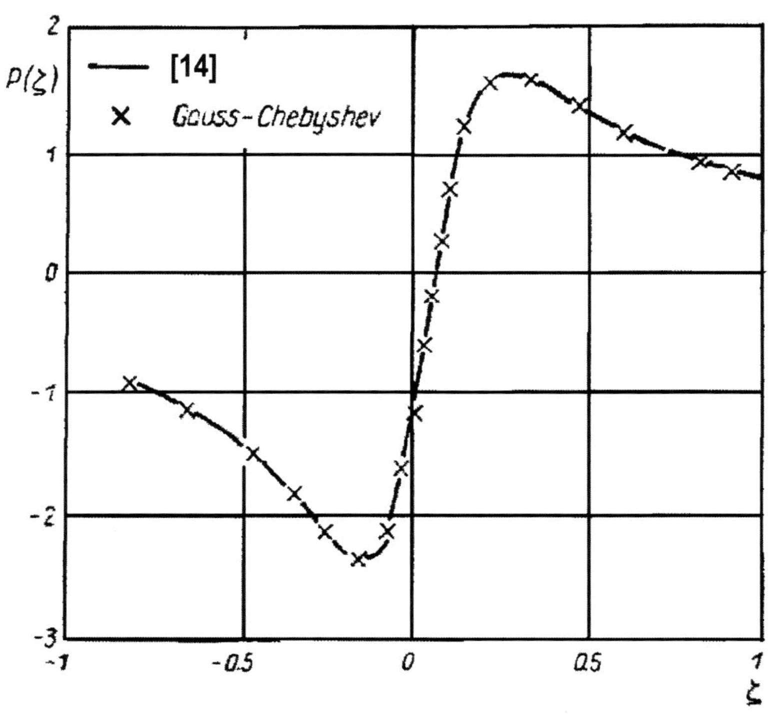

Figure 4.2.3. Pressure distribution $p(\zeta)$ for downwash $w(x)/u = 1/(1+25x^2)$ for the planar airfoil of Figure 4.2.1.

As a second application of the airfoil equation, let us consider the following downwash function:

$$\frac{w(x)}{u} = \frac{1}{1+25x^2} \tag{4.2.32}$$

Figure 4.2.3 shows the pressure distribution $p(\zeta)$ for downwash given by eq. (4.2.32).

Finally, as it is easily seen from Figs. (4.2.2) and (4.2.3), the two different numerical rules, the Gauss-Chebyshev and Gauss-Jacobi numerical integration rules coincide very well.

References

[1] V.V. Golubev, *Theory of an aircraft wing of finite span,* Publ. of the Central Aero-hydro-dynamic Institute of Moscow, Moscow-Leningrad (1931).

[2] T. von Karman and J.M. Burgers, General aerodynamic theory-perfect fluid, Vol.2 of *Aerodynamic Theory,* (ed. W.F.Durand), Springer Verlag, Berlin (1936).

[3] H. Schmidt, Strenge Lösungen zur Prandtlschen Theorie der tragenden Linie, *ZAMM* 17, 101-116 (1937).

[4] K. Schröder, Über eine Integralgleichung erster Art der Tragflügeltheorie, *Sitzungsberichte der Preuss, Akad. d. Wiss. Phys.-nat. Klasse* 30, 345-362 (1938).

[5] K. Schröder, Über die Prandtlsche Integro-differentialgleichung der Tragflügeltheorie, *Abhandl. d. Preuss. Akad. d. Wiss. Math. Naturwiss. Klasse* 16, (1939).

[6] J. Weissinger, Ein Satz über Fourierreichen und seine Anwendung auf die Tragflügeltheorie, *Math. Zeitschr.* 47, 16-33 (1940).

[7] H. Küssner and L. Schwartz, Der schwingende Flügel mit aerodynamisch ausgeglichengem Ruder, *Luftfahrtforschung* 17, 337-354 (1949).

[8] L.G. Magnaradze, On a new integral equation of the theory of aircraft wings, *Soob. A. N. Gruz. SSR* 3, 503-508 (1942).

[9] L.G. Magnaradze, On a system of linear singular integro-differential equations and on the linear Riemann boundary problem, *Soob. A. N. Gruz. SSR* 4, 3-9 (1943).

[10] L.G. Magnaradze, The theory of a class of linear singular integro-differential equations and its application to the problem of vibration of an aircraft wing of finite span, *Soob. A. N. Gruz. SSR* 4, 103-110 (1943).

[11] I.N. Vekua, On Prandtl's integro-differential equation, *Prikl. Mat. i Mech.* 9, 143-150 (1945).

[12] I.N. Vekua, Allgemeine Darstellung der lösungen elliptischer Differentialgleichungen in einem mehrfach Zusammenhängenden gebiet, *Soob. A. N. Gruz. SSR* 1, 329-335 (1940).

[13] H. Schöngen, Die Lösung der Integralgleichung $g(x) = 1/2\pi i \int_{-a}^{a} \frac{f(\xi)d\xi}{x-\xi}$ und deren Anwendung in der Tragflügeltheorie, *Math. Zeitschr.* 45, 245-264 (1939).

[14] S.R. Bland, The two-dimensional oscillating airfoil in a wind tunnel in subsonic flow, *SIAM J. Appl. Math.* 18, 830-848 (1970).

[15] S.R. Bland, R.H.Rhyne and H.B.Pierce, A study of flow-induced vibrations of a plate in narrow channels, *Trans. ASME, Serie B* 89, 824-830 (1967).

[16] J. Blackwell and G. Pounds, Wind-tunnel wall interference effects on a supercritical airfoil at transonic speeds, *J. Aircr.* 14, 929-935 (1977).

[17] J.A. Fromme and M.A. Golberg, Unsteady two-dimensional airloads acting on oscillating thin airfoils in subsonic ventilated wind tunnels, *NASA CR 2967,* Washington (1978).

[18] J.A. Fromme and M.A. Golberg, Numerical solution of a class of integral equations in two-dimensional aerodynamics - the problem of flaps, in *Solution Methods for Integral Equations, Theory and Applications,* (ed. M.A.Golberg), Plenum Press, New York (1979).

[19] J.A. Fromme and M.A. Golberg, Aerodynamic interference effects on oscillating airfoils with controls in ventilated wind tunnels, *AIAA J.* 78, 417-426 (1980).

[20] J.A. Fromme and M.A. Golberg, Reformulation of Possio's kernel with application to unsteady wind tunnel interference, *AIAA J.* 18, 951-957 (1980).

[21] J.A. Fromme and M.A.Golberg, Convergence and stability of a collocation method for the generalized airfoil equation, *Appl. Math. Comp.* 8, 281-292 (1981).

[22] M.A. Golberg and J.A. Fromme, On the L_2 convergence of collocation for the generalized airfoil equation, *J. Math. Anal. Appl.* 71, 271-286 (1979).

[23] M.A. Golberg, M. Lea and G. Miel, A superconvergence result for the generalized airfoil equation with application to the flap problem, *J. Int. Eq.* 5, 175-185 (1983).

[24] M.A. Golberg, The numerical solution of Cauchy singular integral equations with constant coefficients, *J. Int. Eq.* 9, 127-151 (1985).

[25] E.G. Ladopoulos, Finite-part singular integro-differential equations arising in two-dimensional aerodynamics, *Arch. Mech.* 41, 925-936 (1989).

[26] W.F. Moss, The two-dimensional oscillating airfoil a new implementation of the Galerkin method, *SIAM J. Num. Anal.* 20, 391-399 (1983).

[27] W.F. Moss, Numerical solution of integral equations with convolution kernels, *J. Int. Eq.* 4, 253-264 (1982).

[28] D.J. Salmond, Evaluation of two-dimensional subsonic oscillatory airforce coefficients and loading distributions, *Aeronaut. Quart.* 32, 199-211 (1981).

[29] M.H. Williams, The resolvent of singular integral equations, *Q. Appl. Math.* 35, 99-110 (1977).

[30] E. Kraft and C.Lo, Analytical determination of blockage effects in a perforated wall transonic wind tunnel, *AIAA J.* 15, 511-516 (1977).

[31] M. Mokry, Integral equation method for subsonic flow past airfoils in ventilated wind tunnels, *AIAA J.* 13, 47-53 (1975).

[32] E. Nissim and I. Lottati, Oscillatory subsonic piecewise continuous kernel function method, *J. Aircraft* 14, 515-516 (1977).

Chapter 5

COMPUTATIONAL RECIPES OF MULTIDIMENSIONAL SINGULAR INTEGRAL EQUATIONS

5.1. INTRODUCTION

Applications of the multidimensional singular integral equations are given in many fields of engineering mechanics and mathematical physics, like elasticity, plasticity, thermoelastoplasticity, viscoelasticity, viscoplasticity and fracture mechanics theory. F.G. Tricomi [1], [2] was the first scientist who wrote the first important work on multidimensional singular integrals and investigated double singular integrals of the following form:

$$I(x_o, y_o) = \int_S w(x,y) \frac{f(x_o, y_o, \theta)}{r^2} u(x,y)\, dS \tag{5.1.1}$$

where: $(x - x_o) + i\,(y - y_o) = r\,e^{i\theta}$ (5.1.2), and $w(x,y)$ is the weight function for various quadrature rules.

In eqn. (5.1.1), if the density function $u(x_o, y_o)$ is a bounded and Hölder-continuous function in S and also if the characteristic function $f(x_o, y_o\ \theta)$ is bounded and for a fixed pole $X(x_o, y_o)$ is continuous with respect to θ, then according to Tricomi the necessary and sufficient condition for the existence of the singular integral I in its principal value sense, is that its characteristic should satisfy the following condition :

$$\int_0^{2\pi} f(x,y,\theta)\, d\theta = 0 \tag{5.1.3}$$

The multidimensional singular integrals were further studied by G. Giraud [3]-[5]. He investigated integrals taken over a closed Lyapounov manifold Γ of any dimension m.

This manifold is broken up into a finite number of mutually overlapping parts, each one of which has a one-to-one mapping on a region of an m - dimensional Euclidean space.

Giraud investigated singular integrals of the following form:

$$\int_{\Gamma} K(x,y)\, u(y) d\Gamma_y \tag{5.1.4}$$

where the function $u(y)$ satisfies a Lipschitz condition with positive index.

He further studied the following singular integral equation:

$$u(x) - \mu \int_{\Gamma} K(x,y)\, u(y)\, d\Gamma_y = f(x) \tag{5.1.5}$$

with a kernel, the singular part $K_I\,(x,y)$ of which has the following special form:

$$K_I\,(x,y) = \sum_{a=1}^{m} C_a(x)\ (x_a - y_a)\,[\ \sum_{\beta,\gamma=1}^{m} A_{\beta\gamma}\ (x_\beta - y_\beta)(x_\gamma - y_\gamma)\,]^{\frac{-m+1}{2}} \tag{5.1.6}$$

in which $C_a\,(x)$ are certain given functions.

By operating on both sides of eqn. (5.1.5) with:

$$u(x) + \mu \int_{\Gamma} H(x,y,\mu)\, u(y) \tag{5.1.7}$$

where $H(x,y,\mu)$ is any singular kernel, then we shall get the following equation:

$$[\,I + \mu^2\, \Phi(x,\mu)\,]\, u(x) + \mu \int_{\Gamma} [H(x,y,\mu) - K(x,y) - \mu \int_{\Gamma} H(x,z,\mu)\, K(z,y)\, d\Gamma_z\,]\, u(y)\, d\Gamma_y$$

$$= f(x) + \mu \int_{\Gamma} H(x,y,\mu)\, f(y)\, d\Gamma_y \tag{5.1.8}$$

where $\Phi(x,\mu)$ is a function, completely determined by the kernels K and H and the manifold Γ.

Further investigations were done by S.G. Mikhlin [6] - [9] who proved that a singular operator of the type:

$$a\, u(x) + \int_{E_m} \frac{f(\theta)}{r^m}\, u(y)\, dy \tag{5.1.9}$$

where: $r = |y - x|$ and: $\theta = \dfrac{y-x}{r}$ (5.1.10)

(E_m is an Euclidean space of m dimensions) is bounded in the Hilbert - Euclidean space $L_2(E_m)$ if the symbol of the operator is bounded and its norm doesn't exceed the maximum of the modulus of the symbol.

Moreover, S.G. Mikhlin has studied the following integral equation:

$$a(x)\, u(x) + \int_{E_m} \frac{f(x,\theta)}{r^m} u(y)\, dy \tag{5.1.11}$$

On the other hand, if the symbol satisfies certain demands with regard to smoothness, then a multidimensional singular integral equation permits regularization, if and only if, the modulus of its symbol has a positive lower band. The theorem about the regularization can be extended also to systems of singular equations and likewise for the case where the singular integral entering the equation is taken over any closed Lyapounov manifold.

Beyond the above, the basic problem studied by A.P. Calderon and A. Zygmound [10] - [13] are singular integrals of the form:

$$\int_{E_m} \frac{f(\theta)}{r^m}\, u(y)\, dy \tag{5.1.12}$$

They investigated integral (5.1.12) in the Lebesque-Euclidean spaces $L_p(E_m)$ where $1 < p < \infty$, $p \neq 2$. They proved that the integral (5.1.12) is bounded in $L_p(E_m)$ if $f(\theta)$ satisfies the Dini condition :

$$\int_0^1 \frac{\omega(t)}{t}\, dt < \infty \tag{5.1.13}$$

where $\omega(t)$ is the modulus of continuity of the characteristic $f(\theta)$. Also, A.P. Calderon and A. Zygmound investigated the compounding of singular operators of the type:

$$K\, u = a\, u(x) + \int_{E_m} \frac{f(\theta)}{r^m}\, u(y)\, dy \tag{5.1.14}$$

On the other hand, V.D. Kupradze [14], has used multidimensional singular integral equations for the solution of three-dimensional problems of elasticity and thermoelasticity. Moreover, P.P. Zabreyko [15] in his well-known monograph proved some basic theorems on multidimensional singular integral equations.

During the last three decades the Boundary Element Method (B.E.M.), or Boundary Integral Equation Method (B.I.E.M.), has been introduced and investigated for the numerical solution of the multidimensional singular integral equations in combination with some important problems of two- and three-dimensional thermoelastoplasticity, viscoelasticity, viscoplasticity, fracture mechanics and elastodynamics. The Boundary Element Method has been investigated and applied in several problems of engineering mechanics by several scientists. Among them we shall mention the following: J.D. Achenbach, N. Nishimura and J.C. Sung [16], M.H. Aliabadi, D.P. Rooke and D. J. Cartwright [17], N. Altiero and D. Sikarskie [18], J. Balas, V. Sladek and J. Sladek [19], P. K. Banerjee [20], J. Batista de Paiva [21], G. Bezine [22], C. A. Brebbia [23] - [25], C. A. Brebbia, J. C. F. Telles and L. C. Wrobel [26], H. D. Bui [27], J. A. M. Carrer and W. J. Mansur [28],[29], A. Chandra and S. Mukherjee [30], H. B. Chen, P. Lu, M. G. Huang and F. W. Williams [31], W. H. Chen and T. C. Chen [32], S. Christiansen and E. Hansen [33], T. A. Cruse [34], T. A. Cruse and W. Vanburen [35], R. P. Gilbert, G. C. Hsiao and M. Schneider [36], R. P. Gilbert and R. Magnanini [37], R. P. Gilbert and M. Schneider [38], L. J. Gray, D. Ghosh and T. Kaplan [39], M. Guiggiani, G. Krishnasamy, T. J. Rudolphi and F. J. Rizzo [40], G. A. Hartley and A. Abdel-Akher [41], G. A. Hartley [42], M. Heinlein, S. Mukherjee and O. Richmond [43], U. Heise [44], J. C. Lachat [45], V. Mantic [46], D. Martin and M. Aliabadi [47], A. Mendelson [48], A. Mendelson and L. U. Albers [49], M. Morjaria and S. Mukherjee [50], S. Mukherjee [51], Y. X. Mukherjee, S. Mukherjee, X. Shi and A. Nagarajan [52], Y. X. Mukherjee, K. Shah and S. Mukherjee [53], A. Nagarajan, S. Mukherjee and E. Lutz [54], K. S. Parihar and S. Sowdamini [55], F. Paris and S. de León [56], H. Poon, S. Mukherjee and M. F. Ahmad [57], N. N. V. Prasad, M. H. Aliabadi and D. P. Rooke [58]-[59], Y. F. Rashed and M. H. Aliabadi [60], Y. F. Rashed, M. H. Aliabadi and C. A. Brebbia [61], F. J. Rizzo and D. J. Shippy [62], M. A. Sales and L. J. Gray [63], D. Segond and A. Tafreshi [64], V. Sladek, J. Sladek and M. Tanaka [65], M. A. Sutton, C. H. Liu, J. R. Dickerson and S. R. McNeill [66], J. L. Swedlow and T. A. Cruse [67], M. Tanaka, V. Sladek and J. Sladek [68], A. N. Teong and D. L. Clements [69], S. M. Vogel and F. J. Rizzo [70], J. Weaver [71], N. Zabaras and S. Mukherjee [72] and others.

Beyond the above, the Singular Integral Operators Method (S.I.O.M.) was proposed by E.G. Ladopoulos [73] - [85] for the numerical evaluation of the multidimensional singular integral equations in combination with the solution of some basic problems of applied mechanics. By the Singular Integral Operators Method are solved problems of two- and three-dimensional elasto-plasticity of isotropic and anisotropic solids, applied problems of linear viscoelasticity and crack problems in isotropic and anisotropic materials [86], [87].

In the present chapter the Singular Integral Operators Method is analysed and investigated. By the above method are solved two- and three-dimensional singular integral equations.

5.2. Numerical Evaluation Methods for Two-Dimensional Singular Integral Equations

The numerical method which is used for the approximation of the two-dimensional singular integral equations and which will be described in the present chapter is known as the Singular Integral Operators Method (S.I.O.M.). [73]-[77], [78]

Let the following two-dimensional singular integral: [73]-[77], [78]

$$I(x_0,y_0)=\int_s w(x,y)\frac{f(x_0,y_0,\Theta)}{r^2}u(x,y)\,\mathrm{d}s \tag{5.2.1}$$

where:

$$(x-x_0)+i(y-y_0)=re^{i\Theta} \tag{5.2.2}$$

and $w(x,y)$ is the weight function for various quadrature rules. Beyond the above, we assume the following assumptions to be valid:

The density function $u(x_0,y_0)$ in eq. (5.2.1) is a bounded and Hölder-continuous function in s, which means that for any given pair of values $(x_{0_1},y_{0_1}),(x_{0_2},y_{0_2})\in s$:

$$\left|u(x_{0_2},y_{0_2})-u(x_{0_1},y_{0_1})\right|\le A\left|x_{0_2}-x_{0_1}\right|^{\mu}+B\left|y_{0_2}-y_{0_1}\right|^{\nu} \tag{5.2.3}$$

with $0<\mu\le 1$, $0<\nu\le 1$ and A, B, constants.

The characteristic function $f(x_0,y_0,\Theta)$ is bounded and for a fixed pole $X(x_0,y_0)$ is continuous with respect to Θ.

Hence, if these two assumptions are valid, then according to Tricomi [2], the necessary and sufficient condition for the existence of the singular integral (5.2.1) in the principal value sense is that its characteristic should satisfy the following condition (see Figure 5.2.1):

$$\int_0^{2\pi} f(x_0,y_0,\Theta)\,\mathrm{d}\Theta=0 \tag{5.2.4}$$

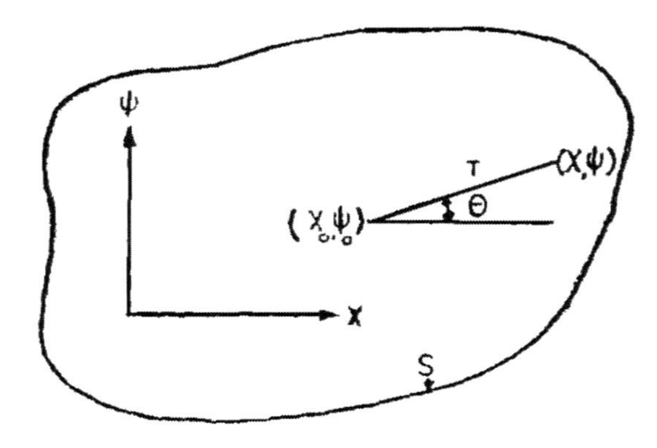

Figure 5.2.1 Geometry of the two-dimensional region *S*.

By assuming the boundary of s to be described by the relation

$$R=R(\Theta),\quad 0<\Theta<2\pi \tag{5.2.5}$$

then eq. (5.2.1) can be further written as:

$$I = \int_{s} \frac{f(\Theta)}{r^2} u(r, \Theta) \, \mathrm{d}s \tag{5.2.6}$$

with:

$$f(\Theta) \equiv f_0(x_0, y_0, \Theta) \tag{5.2.7}$$

and:

$$u(r, \Theta) \equiv u_0(x_0, y_0) \tag{5.2.8}$$

So, from eqs. (5.2.5), (5.2.6), (5.2.7) and (5.2.8), one has:

$$I = \lim_{\varepsilon \to 0} \int_0^{2\pi} f(\Theta) \int_{\varepsilon}^{R(\Theta)} \frac{u(r, \Theta)}{r} \, \mathrm{d}r \, \mathrm{d}\Theta \tag{5.2.9}$$

Furthermore, by using the well-known trapezoidal rule with m abscissae, for the integration with respect to Θ, we consider the following equation:

$$I = \int_0^{2\pi} \Phi(\Theta) \, \mathrm{d}\Theta \cong \frac{2\pi}{m} \sum_{i=0}^{m-1} \Phi\left(\frac{2\pi i}{m}\right) \tag{5.2.10}$$

where:

$$\Phi(\Theta) = f(\Theta) \int_0^{R(\Theta)} \frac{u(r, \Theta)}{r} \, \mathrm{d}r \tag{5.2.11}$$

as it can be seen from (5.2.9).

Moreover, the finite-part integral of eq. (5.2.11) can be evaluated such as:

$$\int_0^{R(\Theta)} \frac{u(r, \Theta)}{r} \, \mathrm{d}r \cong \sum_{k=1}^{n} A_k u(R(\Theta)\rho_k, \Theta) + u(0, \Theta) \ln\left|R(\Theta)\right| \tag{5.2.12}$$

in which ρ_k and A_k are the abscissae and weights.

Thus, from (5.2.11) and (5.2.12) we obtain the numerical solution of (5.2.1):

$$\Phi(\Theta) = f(\Theta)\left[\sum_{k=1}^{n} A_k u(R(\Theta)\rho_k, \Theta) + u(0,\Theta)\ln\left|R(\Theta)\right|\right] \tag{5.2.13}$$

Furthermore, let us use the following integration rules:

i. The Gauss-Legendre Rule

For the Gauss-Legendre numerical integration rule, the weight function takes the form:

$$w(x,y) = 1 \tag{5.2.14}$$

and:

$$f(x_0, y_0, \Theta) = \frac{\vartheta r}{\vartheta x} = \cos\Theta \tag{5.2.15}$$

Then, the two-dimensional integral (5.2.1) becomes:

$$I(x_0, y_0) = \int_s \frac{\vartheta r / \vartheta x}{(x-x_0)^2 + (y-y_0)^2} u(z,y)\,\mathrm{d}s \tag{5.2.16}$$

where s indicates a quadrangular domain with unit half-side length.

On the other hand, by following [74] then the principal value (5.2.1) can be approximated by:

$$I \cong \sum_{l=1}^{m} A_n \frac{[I_A(x_0, y_0, x_m) - I_B(x_0, y_0, x_m)]}{x_m - x_0} - \\ -2[I_A(x_0, y_0, x) - I_B(x_0, y_0, x) K_n(x_0)] \tag{5.2.17}$$

where the functions I_A, I_B are valid as:

$$I_A \cong \sum_{k=1}^{n} A_k \frac{f(x_0, y_k, x)}{y_k - y_0} u(x, y_k) - 2f(x_0, y_0, x) u(x, y_0) K_n(y_0) \\ (\mathit{if}\ y_0 \neq y_k\ (k = 1,2,\ldots,n)) \tag{5.2.18}$$

and:

$$I_B \cong \sum_{\substack{k=1 \\ k \neq m}}^{n} A_k \frac{f(x_0, y_k, x)}{y_k - y_0} u(x, y_k) + A_m \left.\frac{\mathrm{d}[f(x_0, y_0, x) u(x, y)]}{\mathrm{d}y}\right|_{y=y_0} \\ -2f(x_0, y_0, x) u(x, y_0) \Lambda_n(y_0) \quad (\mathit{if}\ y_0 = y_m\ (k = 1,2,\ldots,n)) \tag{5.2.19}$$

with:

$$K_n(x) = \frac{q_n(x)}{\sigma_n(x)} \quad (n_0 \neq n_k) \tag{5.2.20}$$

and:

$$\Lambda_n(x) = \frac{1}{\sigma_t'(x)}\left[q_n'(x) + \frac{1}{4}A_m\sigma_n''(x)\right], \quad (n_0 = n_m) \tag{5.2.21}$$

in which the functions q_n, σ_n are defined according to various numerical integration rules.

Hence, according to eq. (5.2.17), one may write (5.2.18) as:

$$I \cong \int_{-1}^{1} \frac{\mathrm{Im}(I_2(x_0, y_0, y))}{y - y_0} \mathrm{d}y \tag{5.2.22}$$

where $\mathrm{Im}(I_2(x_0, y_0, y))$ denotes the imaginary part of the following complex function:

$$I_2(x_0, y_0, y) = \int_{-1}^{1} u(x,y)\,\mathrm{d}(r/x - z) + \int_{-1}^{1} \frac{r}{(x-z)^2} u(x,y)\,\mathrm{d}x \tag{5.2.23}$$

and:

$$z = x_0 + i(y - y_0) \tag{5.2.24}$$

If y_0 is selected as a root of the Legendre function of the second kind and order m, given by:

$$Q_m(y_0) = 0 \tag{5.2.25}$$

then a general method can be established for the numerical evaluation of the singular integral (5.2.16).

Thus, we finally have:

$$I(x_0, y_0) = \sum_{k=1}^{m} A_k \frac{\mathrm{Im}(I_2(x_0, y_0, y_k))}{(y_k - y_0)} \tag{5.2.26}$$

where y_k are the abscissae and A_k the corresponding weights of the one-dimensional numerical integration rule.

An Application of the Gauss-Legendre Cubature Formulas

Let the following two-dimensional singular integral:

$$I = \int_s \frac{\cos\Theta}{r^2} \mathrm{d}\, s \tag{5.2.27}$$

where s denotes the square $-1 \le x, y \le 1$.

The above integral is solved in closed form as:

$$I = \ln \frac{[1 - y_0 + [(1 + x_0)^2 + (1 - y_0)^2]^{1/2}] \cdot [-1 - y_0 + [(1 - x_0)^2 + (1 + y_0)^2]^{1/2}]}{[-1 - y_0 + [(1 + x_0)^2 + (1 + y_0)^2]^{1/2}] \cdot [1 - y_0 + [(1 - x_0)^2 + (1 - y_0)^2]^{1/2}]} \tag{5.2.28}$$

For the numerical solution of the singular integral (5.2.27), the following numerical integration rule is used, if $f(\Theta)$ satisfies the condition:

$$f(\Theta + \pi) = -f(\Theta) \tag{5.2.29}$$

Then, the use of a numerical integration rule for the evaluation of the singular integrals is simpler than the use of rule (5.2.12).

Thus, for the present case, eq. (5.2.9) can be written as:

$$I = \int_0^{\pi} f(\Theta) \int_{-R(\Theta+\pi)}^{R(\Theta)} \frac{w(r, \Theta)}{r} \mathrm{d}\, r\, \mathrm{d}\, \Theta \tag{5.2.30}$$

where:

$$w(r, \Theta) = \begin{cases} u(r, \Theta), & if\, r > 0 \\ u(-r, \Theta + \pi), & if\, r < 0 \end{cases} \tag{5.2.31}$$

Hence, integral (5.2.30) can be numerically solved as:

$$\int_{-R(\Theta+\pi)}^{R(\Theta)} \frac{w(r, \Theta)}{r} \mathrm{d}\, r \cong$$

$$\cong \sum_{k=1}^{n} A_k \frac{w\left[\dfrac{R(\Theta) + R(\Theta + \pi)}{2}(\rho_k - \delta), \Theta\right]}{\rho_k - \delta} - B(\delta) w(0, \Theta) \tag{5.2.32}$$

where δ is equal to:

$$\delta = -\frac{R(\Theta) - R(\Theta + \pi)}{R(\Theta) + R(\Theta + \pi)} \tag{5.2.33}$$

In eq. (5.2.32), ρ_k and A_k are the abscissae and weights for the regular form of the numerical integration rule and $B(\delta)$ is a new weight coefficient, dependent on the numerical integration rule used.

Beyond the above, the present results may be also used when $u(r,\Theta)$ presents some kind of singularity as $r \to R(\Theta)$ for a fixed Θ.

Therefore, Table 5.2.1 below gives the numerical results by following equation (5.2.10) for m = 18, 36 and 72 and equation (5.2.32) with n = 20 for various values of (x_0,y_0) and the theoretical results by following equation (5.2.28).

Table 5.2.1. Numerical and theoretical results for the singular integral (5.2.27)

m (x_0,y_0)	18	36	72	108	Theoretical values eq.(5.2.28)
(0.4,0.1)	-1.233193	-1.235906	-1.234133	-1.234890	-1.234573
(0.6,0.2)	-2.090208	-2.088043	-2.087700	-2.087719	-2.087723
(0.8,0.4)	-3.418663	-3.420512	-3.419919	-3.419901	-3.419896
(1.2,0.6)	-3.360983	-3.361832	-3.363945	-3.364287	-3.364411
(2.0,1.2)	-1.239982	-1.241372	-1.243823	-1.246027	-1.246424
(2.5,1.5)	-0.430523	-0.431873	-0.433927	-0.434365	-0.434391
(3.0,2.0)	-0.261542	-0.263487	-0.265899	-0.267235	-0.267871
(3.3,2.2)	-0.215434	-0.217853	-0.218934	-0.219325	-0.219426
(4.0,3.2)	-0.120738	-0.121134	-0.121487	-0.121529	-0.121536

ii. The Gauss-Radau rule

The Gauss method described in previous section is used for the solution of open types of singular integrals.

Furthermore, the Radau method is used for the solution of semi-open types and the Lobatto method for the solution of closed types of singular integrals.

In the case of the Radau method, the following quadrature formula is valid:

$$\int_0^1 \frac{u(x)}{x} \mathrm{d}x = \sum_{i=1}^{n} \frac{w_i}{x_i + 1}\left[u\left(\frac{x_i + 1}{2}\right) - u(0)\right] \tag{5.2.34}$$

in which w_i are the weights and x_i denotes a zero of the Legendre polynomial of order n.

By using the trapezoidal rule with m abscissae, and following eqs. (5.2.34) and (5.2.10), the following modified Gauss-Radau cubature formula is obtained:

$$I \cong \frac{2\pi}{m} \times$$

$$\times \sum_{i=1}^{m-1} f\left(\frac{2\pi i}{m}\right)\left[\sum_{j=1}^{n} \frac{w_j}{x_j+1}\left[u\left(\frac{x_j+1}{2}, \frac{2\pi i}{m}\right) - u\left(0, \frac{2\pi i}{m}\right)\right] + u(0, \frac{2\pi i}{m}) \ln\left[R\left(\frac{2\pi i}{m}\right)\right]\right] \tag{5.2.35}$$

where x_j are the zeros of the Legendre polynomials of order n and w_j denotes the weights of the classical Gauss-Legendre quadrature equation.

iii. The Gauss-Lobatto rule

In the case of the Lobatto method, the following quadrature formula is valid:

$$\int_0^1 \frac{u(x)}{x} \mathrm{d}x \cong$$

$$\cong \sum_{i=1}^{n} \frac{w_i}{1-x_i^2}\left[\frac{x_i-1}{2} u(0) - \frac{x_i+1}{2} u(1) + u\left(\frac{x_i+1}{2}\right)\right] - u(0) + u(1) \tag{5.2.36}$$

in which w_i, x_i are the weights and the stations, derived from the special Jacobi polynomials $p^{(1,0)}(x)$.

By using further the trapezoidal rule with m abscissae, and following eqs. (5.2.12) and (5.2.36), then the modified Gauss-Lobatto cubature formula is obtained:

$$I \cong \frac{2\pi}{m} \sum_{i=0}^{m-1} f\left(\frac{2\pi i}{m}\right)\left[\sum_{j=1}^{n} \frac{w_j}{1-x_j^2}\left[\frac{x_j-1}{2} u\left(0, \frac{2\pi i}{m}\right) - \right.\right.$$

$$\left. - \frac{x_j+1}{2} u\left(1, \frac{2\pi i}{m}\right) + u\left(\frac{x_j+1}{2}, \frac{2\pi i}{m}\right)\right] + u\left(1, \frac{2\pi i}{m}\right) +$$

$$\left. + u\left(0, \frac{2\pi i}{m}\right)\left[\ln R\left(\frac{2\pi i}{m}\right)^{-1}\right]\right] \tag{5.2.37}$$

where w_j, x_j are the weights and zeros corresponding to the set of orthogonal polynomials ($p^{(1,0)}(x)$).

5.3. Numerical Evaluation Methods for Three-Dimensional Singular Integral Equations

The Singular Integral Operators Method (S.I.O.M.), as was investigated in the previous section, can be further used for the numerical solution of the three-dimensional singular integral equations. This method is described as follows. [74], [75]

The following integral is a three-dimensional singular integral defined on a three-dimensional finite region V, containing the third-order pole $\left(x_0^1, y^1, z^1\right)$, whose boundary is a closed Lyapounov surface S: [74],[75]

$$I(x^1, y^1, z^1) = \int_V \frac{g(x^1, y^1, z^1, \theta, \varphi)}{r^3} u(x, y, z)\,\mathrm{d}V$$

$$= \int_V \frac{g(x^1, y^1, z^1, \theta, \varphi)}{\left[(x - x^1)^2 + (y - y^1)^2 + (z - z^1)^2\right]^{3/2}} u(x, y, z)\,\mathrm{d}V \tag{5.3.1}$$

Furthermore, we introduce the system of spherical coordinates:

$$x = x^1 + r \sin\theta \cos\varphi$$
$$y = y^1 + r \sin\theta \sin\varphi, \ (0 \le \theta \le \pi) \tag{5.3.2}$$

$$z = z^1 + r\cos\theta, \quad (0 \le \varphi \le 2\pi)$$
$$r^2 = (x - x^1)^2 + (y - y^1)^2 + (z - z^1)^2$$

Then, from eq. (5.3.2) one has:

$$I = \lim_{\varepsilon \to 0} \int_0^{\pi} \int_0^{2\pi} \int_{\varepsilon}^{R(\theta,\varphi)} \sin\theta\, f(\theta, \varphi) \frac{u(r, \theta, \varphi)}{r} \mathrm{d}\theta\, \mathrm{d}\varphi\, \mathrm{d}r \tag{5.3.3}$$

Hence, by integrating eq. (5.3.3) with respect to θ and φ and using the trapezoidal rule with abscissae G, D, then follows the relation:

$$I = \frac{2\pi^2}{GD} \sum_{i=1}^{G} \sum_{j=1}^{D} \sin(\theta_i) \Phi(\theta_i, \varphi_j) \tag{5.3.4}$$

in which:

$$\theta_i = \frac{\pi}{G}(i - 1) \tag{5.3.5}$$
$$\varphi_j = \frac{2\pi(j-1)}{D}$$

and:

$$\Phi(\theta,\varphi) = g(\theta,\varphi) \int_{O}^{R(\theta,\varphi)} \frac{u(r,\theta,\varphi)}{r} \mathrm{d}r \tag{5.3.6}$$

For the numerical evaluation of the integral in (5.3.6) the following integration rule will be used:

$$\int_{0}^{R(\theta,\varphi)} \frac{u(r,\theta,\varphi)}{r} \mathrm{d}r =$$

$$= \sum_{k=1}^{L} A_k u[R(\theta,\varphi)\rho_k, \theta, \varphi] + u(0,\theta,\varphi)\ln[R(\theta,\varphi)] \tag{5.3.7}$$

where ρ_k are the abscissae and A_k the weights for the integration interval [0,1].

REFERENCES

[1] F.G. Tricomi, Formula d'inversione dell'ordine di due integrazioni doppie con asterisco, *Rend. Accad. Naz. Lincei.* 3, 535-539 (1926).

[2] F.G. Tricomi, Equazioni integrali contenti il valor principale di un integrale doppio, *Math. Zeit.* 27, 87-133 (1928).

[3] G. Giraud, Sur differentes questions relatives aux equations du type elliptique, *Ann. Sci Ecole Norm. Sup.* 47, 197-266 (1930).

[4] G. Giraud, Equations a integrales principales, etude suivie d'une application, *Ann.Sci.Ecole Norm. Sup.* 51, 251-372 (1934).

[5] G. Giraud, Equations et systems d'equations on figurent des valeurs principales d'integrales, *C. R.. Acad. Sci.* 204, 628-630 (1937).

[6] S.G. Mikhlin, Compounding of multidimensional singular integrals, *Vestnik Leningr. un-ta* 2, 24-41 (1955).

[7] S.G. Mikhlin, The theory of multidimensional singular integral equations, *Vestnik Leningr.un-ta* 1, 3-24 (1956).

[8] S.G. Mikhlin, Notes on the solutions of multidimensional singular integral equations, *Dokl. Akad. Nauk SSSR* 131, 1019-1021 (1960).

[9] S.G. Mikhlin, *Multidimensional Singular Integrals and Integral Equations*, Pergamon Press, Oxford (1965).

[10] A. Calderon and A. Zygmound, On the existence of certain singular integrals, *Acta Math.* 88, 85-139 (1952).

[11] A. Calderon and A. Zygmound, On the singular integrals, *Amer. J. Math.* 78, 289-309 (1956).

[12] A. Calderon and A. Zygmound, Singular integral operators and differential equations, *Amer. J. Math.* 79, 901-921 (1957).

[13] A. Calderon, Singular integrals, *Bull. Amer. Math. Soc.* 72, 426-465 (1966).

[14] V.D. Kupradze, *Three-dimensional Problems in the Mathematical Theory of Elasticity and Thermoelasticity*, Nauka, Moscow (1976).

[15] P.P. Zabreyko, *Integral Equations-A Reference Text,* Noordhoff, Leyden, The Netherlands (1975).

[16] J.D. Achenbach, N. Nishimura and J.C. Sung, Crack - tip fields in a viscoplastic material, *Int J. Solid. Struct.* 23, 1035-1052 (1987).

[17] M.H. Aliabadi, D.P. Rooke and D.J. Cartwright, An improved boundary element formulation for calculating stress intensity factors: Application to aerospace structures, *J. Str. Anal.* 22, 203-207 (1987).

[18] N. Altiero and D. Sikarskie, A boundary integral method applied to plates of arbitrary plane form, *Comp. Struct.* 9, 163-168 (1978).

[19] J. Balas, V. Sladek and J. Sladek, The boundary integral equation method for plates resting on a two parameter foundation, *ZAMM* 51, 574-580 (1984).

[20] P.K. Banerjee, *The Boundary Element Methods in Engineering,* McGraw-Hill, Berkshire (1994).

[21] J. Batista de Paiva, Boundary element formulation of building slabs, *Engng Anal. Bound. Elem.* 17, 105-110 (1996).

[22] G. Bezine, Boundary integral equations for plate flexure with arbitrary boundary conditions, *Mech. Res. Comm.* 5, 197-206 (1978).

[23] C.A. Brebbia, *The Boundary Element Method for Engineers,* Pentech Press, London (1978).

[24] C.A. Brebbia, *The Boundary Element Method for Engineers,* 2nd revised ed., Pentech Press, London (1980).

[25] C.A. Brebbia, *Progresses in the Boundary Element Method*, Pentech Press, London (1981).

[26] C.A. Brebbia, J.C.F.Telles and L.C.Wrobel, *Boundary Element Techniques : Theory and Applications in Engineering,* Springer-Verlag, Berlin (1984).

[27] H.D. Bui, Some remarks about the formulation of three-dimensional thermoelastoplastic problems of integral equations, *Int. J. Sol. Struct.* 14, 935-939 (1978).

[28] J.A.M. Carrer and W.J. Mansur, Time-domain BEM analysis for the 2D scalar wave equation: initial conditions contributions to space and time derivatives, *Int. J. Numer. Meth. Engng* 39, 2169-2188 (1996).

[29] J.A.M. Carrer and W.J. Mansur, Stress and velocity in 2D transient elastodynamic analysis by the boundary element method, *Engng Anal. Bound. Elemen.* 23, 233-245 (1999).

[30] A. Chandra and S. Mukherjee, A boundary element formulation for large strain problems of compressible plasticity, *Engng Anal.* 3, 71-78 (1986).

[31] H.B. Chen, P. Lu, M.G. Huang and F.W. Williams, An effective method for finding values on and near boundaries in the elastic BEM, *Comp. Struct.* 69, 421-431 (1998).

[32] W.H. Chen and T.C. Chen, An efficient dual boundary element method technique for a two-dimensional fracture problem with multiple cracks, *Int. J. Numer. Meth. Engng* 38, 1739-1756 (1995).

[33] S. Christiansen and E. Hansen, A direct integral equation method for compounding the hoop stress at holes in plane, isotropic sheets, *J. Elasticity* 5, 1-14 (1975).

[34] T.A. Cruse, Application of the boundary-integral equation method to three-dimensional stress analysis, *Comp. Struct.* 3, 509-527 (1973).

[35] T.A. Cruse and W. Vanburen, Three-dimensional elastic stress analysis of a fracture specimen with an edge crack, *J. Fract. Mech.* 7, 1-15 (1971).

[36] R.P. Gilbert, G.C. Hsiao and M. Schneider, The two-dimensional linear orthotropic plate, *Appl. Anal.* 15, 147-169 (1983).

[37] R.P. Gilbert and R. Magnanini, The boundary integral method for two-dimensional orthotropic materials, *J. Elasticity* 18, 61-82 (1987).

[38] R.P. Gilbert and M. Schneider, The linear anisotropic plate, *J. Compos. Mater.* 15, 71-78 (1981).

[39] L.J. Gray, D. Ghosh and T. Kaplan, Evaluation of the anisotropic Green's function in three dimensional elasticity, *Comput. Mech.* 17, 255-261 (1996).

[40] M. Guiggiani, G. Krishnasamy, T.J. Rudolphi and F.J. Rizzo, A general algorithm for the numerical solution of hypersingular boundary integral equations, *ASME J. Appl. Mech.* 59, 604-614 (1992).

[41] G.A. Hartley and A. Abdel-Akher, Analysis of building frames, *ASCE J. Struct. Engng* 119, 468-483 (1993).

[42] G.A. Hartley, Development of plate bending elements for frame analysis, *Engng Anal. Bound. Elem.* 17, 93-104 (1996).

[43] M. Heinlein, S. Mukherjee and O. Richmond, A boundary element method analysis of temperature fields and stress during solidification, *Acta Mech.* 59, 58-81 (1986).

[44] U. Heise, The spectra of some integral operators for plane elastostatic boundary value problems, *J. Elasticity* 8, 47-49 (1978).

[45] J.C. Lachat, A further development of the boundary integral technique for elastostatics, Ph.D. thesis, Southampton University (1975).

[46] V. Mantic, A new formula for the C-matrix in the Somigliana identity, *J.Elasticity* 33, 191-201 (1993).

[47] D. Martin and M. Aliabadi, A BE hyper-singular formulation for contact problems using non-conforming discretization, *Comp. Struct.* 69, 557-565 (1998).

[48] A. Mendelson, Boundary integral methods in elasticity and plasticity, *Report No. NASA TND-7418* (1973).

[49] A. Mendelson and L.U. Albers, Application of boundary-integral equations to elastoplastic problems, in *Boundary Integral Equation Method: Computational Applications in Applied Mechanics,* (eds: T.A.Cruse and F.J.Rizzo), p.p.47-84, ASME (1975).

[50] M. Morjaria and S. Mukherjee, Numerical analysis of planar, time-dependent inelastic deformation of plates with cracks by the boundary element method, *Int. J. Sol. Struct.* 17, 127-143 (1981).

[51] S. Mukherjee, Corrected boundary integral equation in planar thermoelastoplasticity, *Int. J . Sol. Struct.* 13, 331-335 (1977).

[52] Y.X. Mukherjee, S.Mukherjee, X.Shi and A.Nagarajan, The boundary contour method for three-dimensional linear elasticity with a new quadratic boundary element, *Engng Anal. Bound. Elem.* 20, 35-44 (1997).

[53] Y.X. Mukherjee, K. Shah and S. Mukherjee, Thermoelastic fracture mechanics with regularized hypersingular boundary integral equations, *Engng Anal. Bound. Elem.* 23, 89-96 (1999).

[54] A. Nagarajan, S. Mukherjee and E. Lutz, The boundary contour method for three dimensional linear elasticity, *ASME J. Appl. Mech.* 63, 278-286 (1996).
[55] K.S. Parihar and S. Sowdamini, Stress distribution in a two-dimensional infinite anisotropic medium with collinear cracks, *J. Elasticity* 15, 193-214 (1985).
[56] F. Paris and S.de León, Thin plates by the boundary element method by means of two Poisson equations, *Engng Anal. Bound. Elem.* 17, 111-122 (1996).
[57] H. Poon, S. Mukherjee and M.F. Ahmad, Use of "simple solutions" in regularizing hypersingular boundary integral equations in elastoplasticity, *ASME J. Appl. Mech.* 65, 39-45 (1998).
[58] N.N.V. Prasad, M.H. Aliabadi and D.P. Rooke, The dual boundary element method for thermoelastic crack problems, *Int. J. Fract.* 66, 255-272 (1994).
[59] N.N.V. Prasad, M.H. Aliabadi and D.P. Rooke, The dual boundary element method for transient thermoelastic crack problems, *Int. J. Solids Struct.* 33, 2695-2718 (1996).
[60] Y.F. Rashed and M.H. Aliabadi, Fundamental solutions for thick foundation plates, *Mech. Res. Commun.* 24, 331-340 (1997).
[61] Y.F. Rashed, M.H. Aliabadi and C.A. Brebbia, A boundary element formulation for a Reissner plate on a Pasternak foundation, *Comp. Struct.* 70, 515-532 (1999).
[62] J. Rizzo and D.J. Shippy, A method for stress determination in plane anisotropic elastic bodies, *J. Comp. Mater.* 4, 36-61 (1970).
[63] M.A. Sales and L.J. Gray, Evaluation of the anisotropic Green's function and its derivatives, *Comp. Struct.* 69, 247-254 (1998).
[64] D. Segond and A. Tafreshi, Stress analysis of three-dimensional contact problems using the boundary element method, *Engng Anal. Bound. Elem.* 22, 199-214 (1998).
[65] V. Sladek, J. Sladek and M. Tanaka, Evaluation of 1/r integrals in BEM formulations for 3-D problems using coordinate multitransformations, *Engng Anal. Bound. Elem.* 20, 229-244 (1997).
[66] M.A. Sutton, C.H. Liu, J.R. Dickerson and S.R. McNeill, The two-dimensional boundary integral equation method in elasticity with a consistent boundary formulation, *Engng Anal.* 3, 73-84 (1986).
[67] J.L. Swedlow and T.A. Cruse, Formulation of boundary integral equation for three-dimensional elasto-plastic body, *Int. J. Sol. Struct.* 7, 1673-1683 (1971).
[68] M. Tanaka, V. Sladek and J. Sladek, Regularization techniques applied to boundary element methods, *Appl. Mech. Rev.* 47, 457-499 (1994).
[69] A.N. Teong and D.L. Clements, A boundary integral equation method for the solution of a class of crack problems, *J. Elasticity* 17, 9-21 (1987).
[70] S.M. Vogel and F.J. Rizzo, An integral equation formulation of three-dimensional anisotropic elastostatic boundary value problems, *J. Elasticity* 3, 203-216 (1973).
[71] J. Weaver, Three-dimensional crack analysis, *Int. J. Sol. Struct.* 13, 321-330 (1977).
[72] N. Zabaras and S.M. Mukherjee, An analysis of solidification problems by the boundary element method, *Int. J. Num. Meth. Engng* 24, 1879-1900 (1987).
[73] E.G. Ladopoulos, On the numerical evaluation of the singular integral equations used in two and three-dimensional plasticity problems, *Mech. Res. Commun.* 14, 263-274 (1987).
[74] E.G. Ladopoulos, Singular integral representation of three-dimensional plasticity fracture problem, *Theor. Appl. Fract. Mech.* 8, 205-211 (1987).

[75] E.G. Ladopoulos, On the numerical solution of the multidimensional singular integrals and integral equations used in the theory of linear viscoelasticity, *Int. J. Math. Math. Scien.* 11, 561-574 (1988).

[76] E.G. Ladopoulos, Relativistic elastic stress analysis for moving frames, *Rev. Roum. Sci, Tech., Méc. Appl.* 36, 195-209 (1991).

[77] E.G. Ladopoulos, Singular integral operators method for two-dimensional plasticity problems, *Comp. Struct.* 33, 859-865 (1989).

[78] E.G. Ladopoulos, Cubature formulas for singular integral approximations used in three-dimensional elasticity, *Rev. Roum. Sci. Tech., Méc. Appl.* 34, 377-389 (1989).

[79] E.G. Ladopoulos, Singular integral operators method for three-dimensional elasto-plastic stress analysis, *Comp. Struct.* 38, 1-8 (1991).

[80] E.G. Ladopoulos, Singular integral operators method for two-dimensional elasto-plastic stress analysis, *Forsch. Ingen.* 57, 152-158 (1991).

[81] E.G. Ladopoulos, Singular integral operators method for anisotropic elastic stress analysis, *Comp. Struct.* 48, 965-973 (1993).

[82] E.G. Ladopoulos, V.A. Zisis and D. Kravvaritis, Multidimensional singular integral equations in Lp applied to three-dimensional thermoelastoplastic stress analysis, *Comp.Struct.* 52, 781-788 (1994).

[83] E.G. Ladopoulos, *Singular Integral Equations, Linear and Non-linear Theory and its Applications in Science and Engineering,* Springer-Verlag, Berlin, New York (2000).

[84] E.G. Ladopoulos, 3-D elastostatics by coupling method of singular integral equations with finite elements, *Engng Anal. Bound. Elem.* 26, 591-596 (2002).

[85] E.G. Ladopoulos, Coupling of singular integral equation methods and finite elements in 2-D elasticity, *Forsch. Ingen.* 69, 11-16 (2004).

[86] E.G. Ladopoulos and V.A. Zisis, Singular integral representation of two-dimensional shear fracture mechanics problem, *Rev. Roum. Sci. Tech., Méc. Appl.* 38, 617-628 (1993).

[87] V.A. Zisis and E.G. Ladopoulos, Two-dimensional singular integral equations exact solutions, *J. Comput. Appl. Math.* 31, 227-232 (1990).

Chapter 6

ELASTICITY AND FRACTURE MECHANICS OF ISOTROPIC SOLIDS BY MULTIDIMENSIONAL SINGULAR INTEGRAL EQUATIONS

6.1. INTRODUCTION

The theory of multidimensional singular integral equations, with a wide field of applications in engineering mechanics and mathematical physics, such as thermoelastoplasticity, structural analysis and fracture mechanics theory, has developed comparatively slowly over a rather long period. Recently, however, interest has increased sharply.

V.D. Kupradze [1], [2] used multidimensional singular integral equations widely in the solution of three-dimensional elasticity and thermoelasticity problems. On the other hand, over the last decades, the Boundary Element Method (B.E.M.) has been used for the numerical evaluation of multidimensional singular integral equations, applied in elasticity, plasticity, structural analysis and fracture mechanics theory. Among the scientists who used the Boundary Element Method (B.E.M.), or Boundary Integral Equation Method (B.I.E.M.), for the solution of elasticity, structural analysis and fracture mechanics problems of isotropic media we shall mention the following: M. H. Aliabadi, D. P. Rooke and D. J. Cartwright [3], N. Altiero and D. Sikarskie [4], J. Balas, V. Sladek and J. Sladek [5], P. K. Banerjee [6], J. Batista de Paiva [7], G. Bezine [8], C. A. Brebbia [9] - [11], C. A. Brebbia, J. C. F. Telles and L. C. Wrobel [12], H. D. Bui [13], H. B. Chen, P. Lu, M. G. Huang and F. W. Williams [14], W. H. Chen and T. C. Chen [15], S. Christiansen and E. Hansen [16], T. A. Cruse [17], T. A. Cruse and W. Vanburen [18], M. Guiggiani, G. Krishnasamy, T. J. Rudolphi and F. J. Rizzo [19], G. A. Hartley and A. Abdel-Akher [20], G. A. Hartley [21], M. Heinlein, S. Mukherjee and O. Richmond [22], U. Heise [23], J. C. Lachat [24], V. Mantic [25], D. Martin and M. Aliabadi [26], M. Morjaria and S. Mukherjee [27], S. Mukherjee [28], Y. X. Mukherjee, S. Mukherjee, X. Shi and A. Nagarajan [29], Y. X. Mukherjee, K. Shah and S. Mukherjee [30], A. Nagarajan, S. Mukherjee and E. Lutz [31], F. Paris and S. deLeón [32], N. N. V. Prasad, M. H. Aliabadi and D. P. Rooke [33]-[34], Y. F. Rashed and M. H. Aliabadi [35], Y. F. Rashed, M. H. Aliabadi and C. A. Brebbia [36], D. Segond and A. Tafreshi [37], V. Sladek, J. Sladek and M. Tanaka [38], M. A. Sutton, C. H. Liu, J. R. Dickerson and S. R. McNeill [39],

M. Tanaka, V. Sladek and J. Sladek [40], A. N. Teong and D. L. Clements [41], J. Weaver [42], N. Zabaras and S. Mukherjee [43] and others.

In addition, E.G. Ladopoulos [44] - [49] has proposed the Singular Integral Operators Method (S.I.O.M.) for the numerical evaluation of the multidimensional singular integral equations, used in two- and three-dimensional elasticity and viscoelasticity problems and fracture mechanics problems of isotropic solids.

By the present chapter, the following two- and three-dimensional elasticity, viscoelasticity and fracture mechanics problems are solved, by using the Singular Integral Operators Method: Two-dimensional elastic stress analysis of isotropic solids, application to the determination of the stress field in the neighbourhood of a circular hole under internal pressure in an infinite and isotropic solid, three-dimensional elastic stress analysis of isotropic solids, application to the determination of the elastic stress analysis of a thick cylinder subjected to an internal pressure, application to the determination of the stress intensity factors near the edge of a rectangular crack under tensile pressure in an isotropic solid, linear viscoelasticity of isotropic solids and application of linear viscoelasticity.

6.2. Linear Theory of Two-Dimensional Elasticity of Isotropic Solids

The multidimensional singular integral equations are applied in praxis to the solution of several important problems of the general theory of two-dimensional isotropic elasticity. [47]

Consider the following Cauchy-Navier differential equation [42], [47]:

$$L = \left[G\Delta + \frac{G}{1-2v} \operatorname{grad} \operatorname{div} \right] u(x) = -b(x) \tag{6.2.1}$$

where $u(x) = \begin{bmatrix} u_1 \\ u_2 \\ u_3 \end{bmatrix}$ is the displacement vector of a body and $b(x) = \begin{bmatrix} b_1 \\ b_2 \\ b_3 \end{bmatrix}$ is the body force. Furthermore, G is the shear modulus and v Poisson's ratio.

The strain-displacement relations in the linear theory are:

$$\varepsilon_{ij} = \frac{1}{2}\left(\frac{\vartheta u_i}{\vartheta x_j} + \frac{\vartheta u_j}{\vartheta x_i} \right) \tag{6.2.2}$$

and the stress-strain relations:

$$\sigma_{ij} = 2G\varepsilon_{ij} + \frac{2Gv}{1-2v} \Delta_{ij} \varepsilon_{ll} \tag{6.2.3}$$

in which Δ_{ij} is Kronecker's delta, G the shear modulus and v Poisson's ratio. Let by Γ_1 the portion of the boundary of the body on which displacements are prescribed, Γ_2 the surface of the body on which the force tractions are employed and Γ the total surface of the body. Then, the principal virtual displacements is: [10],[17]

$$\int_{\Omega} (\sigma_{jk,j} + b_k) u_k^* \, \mathrm{d}\Omega = \int_{\Gamma_2} (p_k - \bar{p}_k) u_k^* \, \mathrm{d}\Gamma \tag{6.2.4}$$

where u_k^* are the virtual displacements identically satisfying the homogeneous boundary conditions $u_k^* \equiv 0$ on Γ_1.

Eq. (6.2.4) can be further written as:

$$\int_{\Omega} (\sigma_{jk,j} + b_k) u_k^* \, \mathrm{d}\Omega = \int_{\Gamma_2} (p_k - \bar{p}_k) u_k^* \, \mathrm{d}\Gamma + \int_{\Gamma_1} (\bar{u}_k - u_k) p_k^* \, \mathrm{d}\Gamma \tag{6.2.5}$$

in which $p_k^* = n_j \sigma_{jk}^*$ are the surface forces or tractions corresponding to the u_k^* system.

By integrating eq. (6.2.5) one has:

$$\begin{gathered} \int_{\Omega} b_k u_k^* \, \mathrm{d}\Omega - \int_{\Omega} \sigma_{jk} \varepsilon_{jk}^* \, \mathrm{d}\Omega = \\ -\int_{\Gamma_2} \bar{p}_k u_k^* \, \mathrm{d}\Gamma - \int_{\Gamma_1} p_k u_k^* \, \mathrm{d}\Gamma + \int_{\Gamma_1} (\bar{u}_k - u_k) p_k^* \, \mathrm{d}\Gamma \end{gathered} \tag{6.2.6}$$

A second integration of eq. (6.2.6) results:

$$\int_{\Omega} b_k u_k^* \, \mathrm{d}\Omega + \int_{\Omega} \sigma_{jk,j}^* u_k \, \mathrm{d}\Omega = -\int_{\Gamma_2} \bar{p}_k u_k^* \, \mathrm{d}\Gamma - \int_{\Gamma_1} p_k u_k^* \, \mathrm{d}\Gamma + \int_{\Gamma_1} \bar{u}_k p_k^* \, \mathrm{d}\Gamma + \int_{\Gamma_2} u_k p_k^* \, \mathrm{d}\Gamma \tag{6.2.7}$$

By representing unit load at *i* in the "*l*" direction, then the solution is defined as:

$$u_l^i + \int_{\Gamma_1} \bar{u}_k p_k^* \, \mathrm{d}\Gamma + \int_{\Gamma_2} u_k p_k^* \, \mathrm{d}\Gamma = \int_{\Omega} b_k u_k^* \, \mathrm{d}\Omega + \int_{\Gamma_1} p_k u_k^* \, \mathrm{d}\Gamma + \int_{\Gamma_2} \bar{p}_k u_k^* \, \mathrm{d}\Gamma \tag{6.2.8}$$

where u_l^i represents the displacement at *i* in the "*l*" direction.

Generally, if $\Gamma = \Gamma_1 + \Gamma_2$, follows:

$$u_l^i + \int_{\Gamma} u_k p_{lk}^* \, \mathrm{d}\Gamma = \int_{\Gamma} p_k u_k^* \, \mathrm{d}\Gamma + \int_{\Omega} b_k u_k^* \, \mathrm{d}\Omega \tag{6.2.9}$$

By considering unit forces acting in the three directions, then eq. (6.2.9) results:

$$u_l^i + \int_\Gamma u_k p_{lk}^* \, \mathrm{d}\Gamma = \int_\Gamma p_k u_{lk}^* \, \mathrm{d}\Gamma + \int_\Omega b_k u_{lk}^* \, \mathrm{d}\Omega \tag{6.2.10}$$

in which p_{lk}^* and u_{lk}^* represent the surface tractions and displacements in the "k" direction, due to the unit forces acting in the "l" direction.

For a two-dimensional isotropic body the fundamental solution is [47]:

$$u_{lk}^* = \frac{1}{8\pi G(1-v)}\left[(3-4v)\ln\left(\frac{1}{r}\right)\Delta_{lk} + \frac{\vartheta r}{\vartheta x_l}\frac{\vartheta r}{\vartheta x_k}\right] \tag{6.2.11}$$

$$p_{lk}^* = -\frac{1}{4\pi(1-v)r}\left[\frac{\vartheta r}{\vartheta n}\left[(1-2v)\Delta_{kl} + 2\frac{\vartheta r}{\vartheta x_k}\frac{\vartheta r}{\vartheta x_l}\right] - (1-2v)\left(\frac{\vartheta r}{\vartheta x_l}u_k - \frac{\vartheta r}{\vartheta x_k}u_l\right)\right] \tag{6.2.12}$$

where n is a normal to the surface of the body, r the distance from the point of application of the load to point under consideration and n_j the direction cosines, Figure 6.2.1 and 6.2.2.

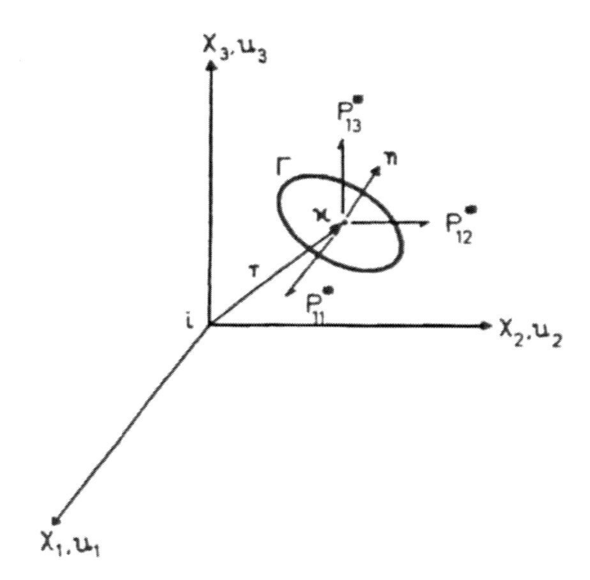

Figure 6.2.1. Surface forces acting at point k, due to unit load at point i acting in the "l" direction.

On the other hand, the displacements at a point for the "l" component appears to be, from (6.2.10):

$$u_l^i = \int_\Gamma u_{lk}^* p_k \, \mathrm{d}\Gamma - \int_\Gamma p_{lk}^* u_k \, \mathrm{d}\Gamma + \int_\Omega b_k u_{lk}^* \, \mathrm{d}\Omega \tag{6.2.13}$$

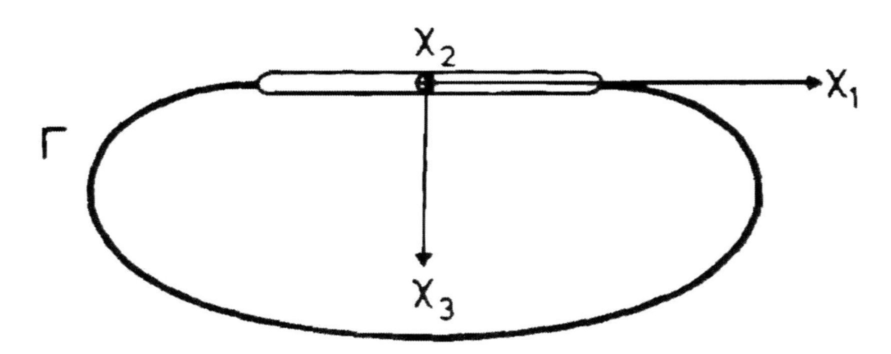

Figure 6.2.2. Geometry of the cracked solid.

Thus, from eq. (6.2.2) and eq. (6.2.3) we obtain the formula for the stresses of an isotropic solid:

$$\sigma_{ij} = G\left(\frac{\vartheta u_i}{\vartheta x_j} + \frac{\vartheta u_j}{\vartheta x_i}\right) + \frac{2G\nu}{1-2\nu}\Delta_{ij}\frac{\vartheta u_i}{\vartheta x_j} \tag{6.2.14}$$

By carrying out the differentiation one has:

$$\sigma_{ij} = \int_\Gamma \left[\frac{2G\nu}{1-2\nu}\Delta_{ij}\frac{\vartheta u_{lk}^*}{\vartheta x_l} + G\left(\frac{\vartheta u_{ik}^*}{\vartheta x_j} + \frac{\vartheta u_{jk}^*}{\vartheta x_i}\right)\right] p_k \,\mathrm{d}\Gamma$$

$$+ \int_\Omega \left[\frac{2G\nu}{1-2\nu}\Delta_{ij}\frac{\vartheta u_{lk}^*}{\vartheta x_l} + G\left(\frac{\vartheta u_{ik}^*}{\vartheta x_j} + \frac{\vartheta u_{jk}^*}{\vartheta x_i}\right)\right] b_k \,\mathrm{d}\Omega \tag{6.2.15}$$

$$- \int_\Gamma \left[\frac{2G\nu}{1-2\nu}\Delta_{ij}\frac{\vartheta p_{lk}^*}{\vartheta x_l} + G\left(\frac{\vartheta p_{ik}^*}{\vartheta x_j} + \frac{\vartheta p_{jk}^*}{\vartheta x_i}\right)\right] u_k \,\mathrm{d}\Gamma$$

Eq. (6.2.15) can be reduced to the following formula:

$$\sigma_{ij} = \int_\Gamma A_{kij} p_k \,\mathrm{d}\Gamma - \int_\Gamma B_{kij} u_k \,\mathrm{d}\Gamma + \int_\Omega A_{kij} b_k \,\mathrm{d}\Omega \tag{6.2.16}$$

where the components A_{kij} and B_{kij} are:

$$A_{kij} = \frac{1}{4\pi(1-\nu)r}[(1-2\nu)(\Delta_{ki} r_{,j} + \Delta_{kj} r_{,i} - \Delta_{ij} r_{,k}) + 2r_{,i} r_{,j} r_{,k}] \tag{6.2.17}$$

$$B_{kij} = \frac{G}{2\pi(1-\nu)r^2}\left\{2\frac{\vartheta r}{\vartheta n}[(1-2\nu)\Delta_{ij} r_{,k} + \nu(\Delta_{ik} r_{,j} + \Delta_{jk} r_{,i}) - 4r_{,i} r_{,j} r_{,k}]\right.$$

$$+2v(n_i r_{,j} r_{,k} + n_j r_{,i} r_{,k}) + (1-2v)(2n_k r_{,i} r_{,j} + n_j \Delta_{ik} + n_i \Delta_{jk}) - (1-4v) n_k \Delta_{ij} \Big\} \qquad (6.2.18)$$

where:

$$r_{,i} = \frac{\vartheta r}{\vartheta x_i}$$

6.3. Two-Dimensional Application to the Determination of the Stress Field in the Neighbourhood of a Circular Hole under Internal Pressure in an Infinite and Isotropic Solid

As an application of the previous theory, the stress field will be determined, in the neighbourhood of a circular hole under internal pressure p in an infinite and isotropic solid, Figure 6.3.1.

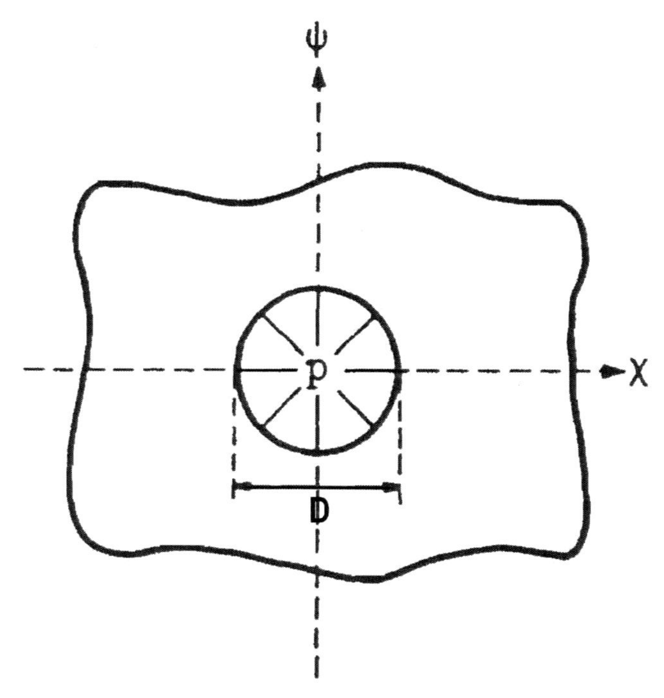

Figure 6.3.1. A circular hole under internal pressure p in an infinite and isotropic solid.

The same problem has been solved theoretically by Muskhelishvilli in his well-known books [50],[51] and a comparison will be made between the theoretical results and the results by using the Singular Integral Operators Method. [47]

The solid used, has the following parameters: Young modulus E=207,900 N/mm^2 and Poisson's ratio v=0.1. The diameter of the circular hole is D=6 mm, Figure 6.3.1.

By using the formula (6.2.16) which gives the stress field in the neighbourhood of an arbitrary shape plane crack, for the circular hole of Figure 6.3.1, it is possible to compute the radial stresses σ_r at internal points of the infinite and isotropic solid.

Therefore, Figure 6.3.2 shows the radial stresses σ_r at internal points of the infinite and isotropic solid at some distances to the center of the circular hole.

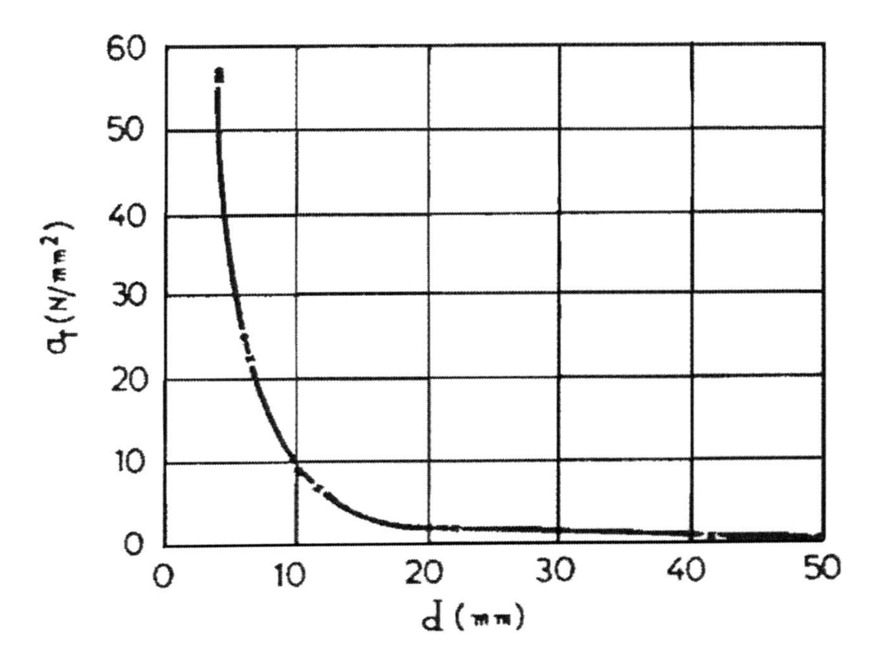

Figure 6.3.2. The radial stresses σ_r at internal points of the infinite and isotropic solid at some distances d to the center of the circular hole.

As it can be seen from Figure 6.3.2, the results of the Singular Integral Operators Method and the theoretical results by using the method of Muskhelishvilli are in fair agreement.

6.4. Linear Theory of Three-Dimensional Elasticity of Isotropic Solids

The multidimensional singular integral equations are further applied in the general theory of three-dimensional elasticity of isotropic solids. [46]

Let us consider an infinite and isotropic elastic solid in three-dimensional space, with Cartesian coordinates ξ_i $(i = 1,2,3)$.

Under arbitrary loadings, the deformation in the solid must be described by the three coordinate variables r, Θ and z and the three displacement components can be found from Navier's equation of equilibrium [17],[18],[46]:

$$\nabla^2 u_r + \frac{1}{1-2v}\frac{\vartheta e}{\vartheta r} - \frac{1}{r^2}\left(2\frac{\vartheta u_\Theta}{\vartheta \Theta} + u_r\right) = 0 \tag{6.4.1}$$

$$\nabla^2 u_\Theta + \frac{1}{1-2v}\frac{1}{r}\frac{\vartheta e}{\vartheta \Theta} + \frac{1}{r^2}\left(2\frac{\vartheta u_r}{\vartheta \Theta} - u_\Theta\right) = 0 \tag{6.4.2}$$

$$\nabla^2 u_z + \frac{1}{1-2v}\frac{\vartheta e}{\vartheta z} = 0 \tag{6.4.3}$$

where v is Poisson's ratio and e is given by the following formula:

$$e = \frac{\vartheta u_r}{\vartheta r} + \frac{u_r}{r} + \frac{1}{r}\frac{\vartheta u_\Theta}{\vartheta \Theta} + \frac{\vartheta u_z}{\vartheta z} \tag{6.4.4}$$

and ∇^2, the Laplacian in three dimensions, by:

$$\nabla^2 = \frac{\vartheta^2}{\vartheta r^2} + \frac{1}{r}\frac{\vartheta}{\vartheta r} + \frac{1}{r^2}\frac{\vartheta^2}{\vartheta \Theta^2} + \frac{\vartheta^2}{\vartheta z^2} \tag{6.4.5}$$

The stress components σ_{ij} and displacement gradients are related by Hook's law:

$$\sigma_{ij} = \frac{2Gv}{1-2v}\delta_{ij}u_{n,n} + G(u_{i,j} + u_{j,i}) \tag{6.4.6}$$

in which v denotes Poisson's ratio, δ_{ij} Kronecker's delta and G the shear modulus.

The corresponding jth component of the displacement is equal to:

$$U_{ij} = \frac{1}{4\pi Gr}\left(\frac{(3-4v)}{4(1-v)}\delta_{ij} + \frac{r_{,i}r_{,j}}{4(1-v)}\right) \tag{6.4.7}$$

Hence, by eq. (6.4.7) is given the solution of Kelvin's problem of the point load in an infinite body.

The jth components of the traction for Kelvin's problem are determined by eq. (6.4.7):

$$T_{ij} = -\frac{K}{r^2}\left[\frac{\vartheta r}{\vartheta n}\left(\delta_{ij} + \frac{3r_{,i}r_{,j}}{1-2v}\right) - n_j r_{,i} + n_i r_{,j}\right] \tag{6.4.8}$$

where:

$$K = (1-2v)/8\pi(1-v) \tag{6.4.9}$$

In eqs. (6.4.7) and (6.4.8), all differentiation is with respect to the field point x:

$$r_{,i} = \frac{\vartheta r}{\vartheta x_i} = \frac{1}{r}(x_i - \xi_i) \tag{6.4.10}$$

On the other hand, the load point p with coordinates ξ_1, ξ_2, ξ_3 is:

$$r = \left[(x_i - \xi_i)(x_i - \xi_i)\right]^{1/2} \tag{6.4.11}$$

Furthermore, we assume that the space S is made by two parts, and we shall always assume the internal region A^+ as finite and the external A^- as connected (see Figure (6.4.1)).

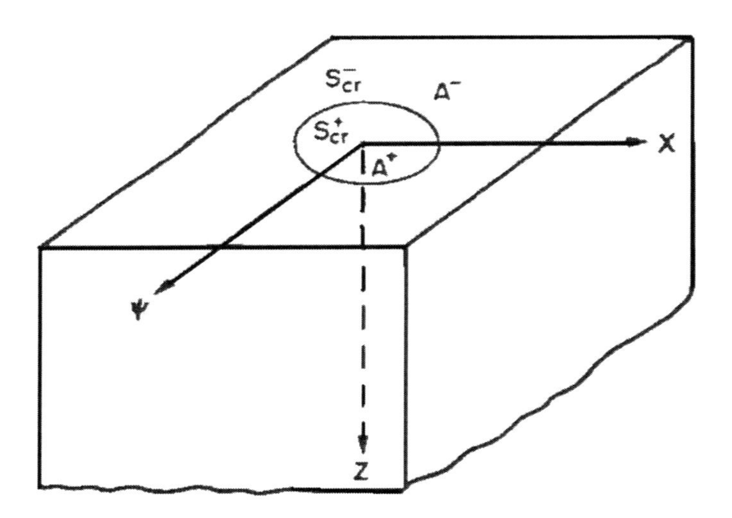

Figure 6.4.1. Geometry of the crack.

We denote further that $S_{cr}^{\pm}$ is a Lyapounov surface, consisting of two planar sheets, the internal one S_{cr}^{+} belonging to the subspace A^+ and the external one denoted by S_{cr}^{-}. It appears that S_{cr}^{+} can be considered as a plane crack.

Betti's third identity can be written as:

$$\int_{S_{cr}^{+}} (u_i T_{ji} - t_i U_{ji})\, \mathrm{d}\, S_{cr}^{+} = 0 \tag{6.4.12}$$

where u_i and t_i are the displacements and traction for the unknown stress state.

Thus, from the application of the Betti-Rayleigh theorem follows:

$$u_i(x) = -\int_{S_{cr}^{+}} u_i(\xi) T_{ij}(\xi, x)\, \mathrm{d}\, S(\xi) + \int_{S_{cr}^{+}} t_j(\xi) U_{ij}(\xi, x)\, \mathrm{d}\, S(\xi) \tag{6.4.13}$$

for points x inside A, where the vectors $u_i(\xi)$ and $t_j(\xi)$ are the displacements and tractions on the surface of the body. Hence, eq. (6.4.13) defines Somigliana's identity for the displacements inside the body.

For the components of the transformed stress tensor, the following identity is valid:

$$\sigma_{ij}(x) = -\int_{S_{cr}^+} u_k(\xi) B_{kij}(\xi, x)\, \mathrm{d}S(\xi) + \int_{S_{cr}^+} t_k(x) C_{kij}(\xi, x)\, \mathrm{d}S(\xi) \tag{6.4.14}$$

with:

$$C_{kij} = \frac{K}{r^2}\left(\delta_{ki} r_{,j} + \delta_{kj} r_{,i} - \delta_{ij} r_{,k} + \frac{3}{1-2v} r_{,i} r_{,j} r_{,k}\right) \tag{6.4.15}$$

and:

$$B_{kij} = \frac{2KG}{r^3}\left[3\frac{\vartheta r}{\vartheta n}\left(\delta_{ij} r_{,k} + \frac{v}{1-2v}(\delta_{ki} r_{,k} + \delta_{kj} r_{,i}) - \frac{5}{1-2v} r_{,i} r_{,j} r_{,k}\right)\right.$$
$$\left. + \frac{3v}{1-2v}(n_i r_{,j} r_{,k} + n_j r_{,i} r_{,k}) + 3n_k r_{,i} r_{,j} + n_j \delta_{ki} + n_i \delta_{kj} - \frac{1-4v}{1-2v} n_k \delta_{ij}\right] \tag{6.4.16}$$

From eqs. (6.4.6) and (6.4.14) we obtain:

$$\sigma_{i3} = G(u_{i,3}^+ + u_{3,i}^+) \quad \textit{for } i = 1,2 \tag{6.4.17}$$

and:

$$\sigma_{33} = 2G[v u_{i,i}^+ + (1-v) u_{3,3}^+]/(1-2v) \tag{6.4.18}$$

In addition, $T_{ij}^+ = T_{ij}^-$ and $U_{ij}^+ = U_{ij}^-$, thus eq. (6.4.13) can be written as:

$$u_i(x) = -\int_{S_{cr}^+} \Delta u_j(\xi) T_{ij}^+(\xi, x)\, \mathrm{d}S(\xi) \tag{6.4.19}$$

Thus, from eqs. (6.4.17) - (6.4.19) we have:

$$\sigma_{i3} = \frac{E(1-2v)}{16\pi(1-v^2)} \times \int_{S_{cr}^+}\left(\delta_{bi} r_{,c} - \delta_{ic} r_{,b} + \frac{3r_{,i} r_{,b} r_{,c}}{1-2v}\right)\frac{\Delta u_{b,c}\, \mathrm{d}S}{r^2} \tag{6.4.20}$$

for $i = 1,2$ and:

$$\sigma_{33} = \frac{E}{8\pi(1-v^2)} \int_{S_{cr}^+} \frac{r_{,i} \Delta u_{3,i}\, \mathrm{d}S}{r^2} \tag{6.4.21}$$

where E denotes Young's modulus.

Therefore, eqs. (6.4.20) and (6.4.21) denote the general three-dimensional elastic stress analysis formulation in an infinite and isotropic solid.

6.5. THREE-DIMENSIONAL APPLICATION TO THE DETERMINATION OF THE STRESS FIELD OF A THICK CYLINDER UNDER AN INTERNAL PRESSURE

For the solution of the surface integrals, which appear in eqs. (6.4.20) and (6.4.21), we introduce the following numerical method [44] - [46]:

Let us consider the surface integral:

$$I(x_0, y_0) = \int_s w(x,y) \frac{f(x_0, y_0, \Theta)}{r^2} u(x,y) \, \mathrm{d}s \tag{6.5.1}$$

where:

$$(x - x_0) + i(y - y_0) = re^{i\Theta} \tag{6.5.2}$$

and $w(x,y)$ denotes the weight function, which in this case can be written as follows due to eq. (6.4.20):

$$w(x,y) = \delta_{bi} r_{,c} - \delta_{ic} r_{,b} + \frac{3 r_{,i} r_{,b} r_{,c}}{1 - 2v} \tag{6.5.3}$$

and as follows due to eq. (6.4.21):

$$w(x,y) = r_{,i} \tag{6.5.4}$$

The Gauss-Legendre cubature formulas

By using the trapezoidal rule with m abscissae, which is applied with respect to Θ, then eq. (6.5.1) takes the form:

$$I \cong \frac{2\pi}{m} \sum_{i=0}^{m-1} \Phi\left(\frac{2\pi i}{m}\right) \tag{6.5.5}$$

in which:

$$\Phi(\Theta) = f(\Theta) \times \left[\sum_{k=1}^{n} A_k u(R(\Theta)\rho_k, \Theta) + u(\rho, \Theta) \ln\left|R(\Theta)\right| \right] \tag{6.5.6}$$

where ρ_k are the abscissae and A_k the weights, with density function $u(x,y)$.

For the Gauss-Legendre numerical integration rule, the weight function is:

$$w(x,y)=1 \tag{6.5.7}$$

and furthermore is valid:

$$f(x_0,y_0,\Theta)=\frac{\vartheta r}{\vartheta x}=\cos\Theta \tag{6.5.8}$$

Therefore, the principal value of eq. (6.5.5) can be approximated by:

$$I\cong\sum_{l=1}^{m}A_n\frac{[I_A(x_0,y_0,x_m)-I_B(x_0,y_0,x_m)]}{x_m-x_0}-2\big(I_A(x_0,y_0)-I_B(x_0,y_0)k_n(x_0)\big) \tag{6.5.9}$$

where the functions I_A and I_B are given by the relations:

$$I_A=\sum_{k=1}^{n}A_k\frac{f(x_0,y_0,x,y_k)}{y_k-y_0}u(x,y_k)-2f(x_0,y_0,x)u(x,y_0)K_n(y_0) \tag{6.5.10}$$

$(if\; y_0\neq y_k, k=1,2,\ldots,n)$ and:

$$I_B\cong\sum_{\substack{k=1\\k\neq m}}^{n}A_k\frac{f(x_0,y_0,x,y_k)}{y_k-y_0}u(x,y_k)+A_m\frac{\mathrm{d}[f(x_0,y_0,x,y)u(x,y)]}{\mathrm{d}y}\bigg|_{y=y_0}$$

$$-2f(x_0,y_0,x)u(x,y_0)\Lambda_n(y_0) \tag{6.5.11}$$

$(if\; y_0=y_m, k=1,2,\ldots,n)$ with:

$$K_n(x)=\frac{q_n(x)}{\sigma_n(x)}\quad(n_0\neq n_k) \tag{6.5.12}$$

and:

$$\Lambda_n(x)=\frac{1}{\sigma_i'(x)}\left[q_n'(x)+\frac{1}{4}A_m\sigma_n''(x)\right],\quad(n_0=n_m) \tag{6.5.13}$$

Beyond the above, eq. (6.5.5) can be written as:

$$I\cong\int_{-1}^{1}\frac{\mathrm{Im}(I_2(x_0,y_0,x))}{y-y_0}\mathrm{d}y \tag{6.5.14}$$

where $\mathrm{Im}(I_2(x_0,y_0,x))$ denotes the imaginary part of the complex function:

$$I_2(x_0, y_0, x) = \int_{-1}^{1} u(x,y)\,\mathrm{d}(r/x - z) + \int_{-1}^{1} \frac{r}{(x-z)^2} u(x,y)\,\mathrm{d}x \tag{6.5.15}$$

with:

$$z = x_0 + i(y - y_0) \tag{6.5.16}$$

Finally, one has:

$$I(x_0, y_0) = \sum_{k=1}^{m} A_k \frac{\mathrm{Im}(I_2(x_0, y_0, y_k))}{(y_k - y_0)} \tag{6.5.17}$$

where n_k are the abscissae and A_k the corresponding weights of the one-dimensional numerical integration rule.

As an application of the previous theory let us determine the stress field of a thick cylinder subjected to an internal pressure p (Figure 6.5.1). The same problem has been numerically evaluated by using finite elements and boundary elements by Lachat [24], and theoretically solved by the method of Muskhelishvilli [51].

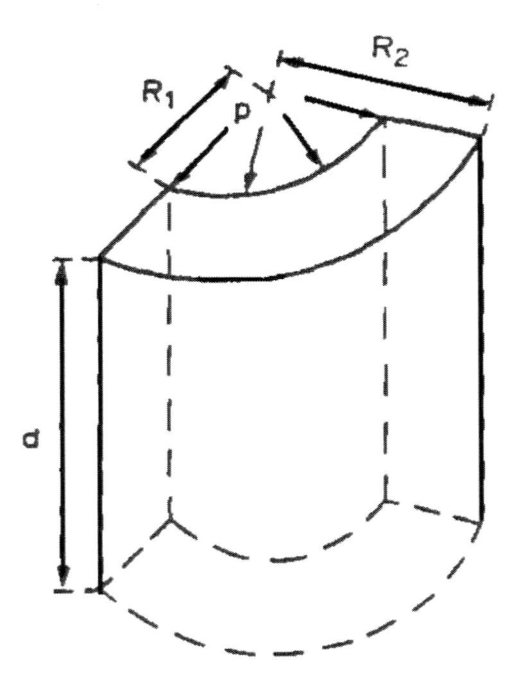

Figure 6.5.1. A thick cylinder subjected to an internal pressure p.

Thus, a comparison will be made between the theoretical results and the numerical results by using the Singular Integral Operators Method (SIOM), the Finite Element Method (FEM) and the Boundary Element Method (B.E.M.). The solid used has the following parameters:

Young's modulus $E = 2.1\times10^5$ N/mm^2; Poisson's ratio $v = 0.3$; internal radius of the cylinder $R_1 = 10$ mm; external radius $R_2 = 20$ mm; and length of the cylinder $d = 40$ mm (see Figure 6.5.1). Moreover, the cylinder is subjected to an internal pressure $p = 20$ N/mm^2.

The radial displacements, and the radial and hoop stresses have been computed by using the corresponding formulas of Section 6.4. Furthermore, the previous mentioned numerical methods have been used for solving the formulas of the displacement and stress field.

Hence, Figure 6.5.2 shows the relation between the radial displacements u_r and the radius R of the thick cylinder, by using theoretical methods and the numerical methods of the SIOM, FEM and BEM.

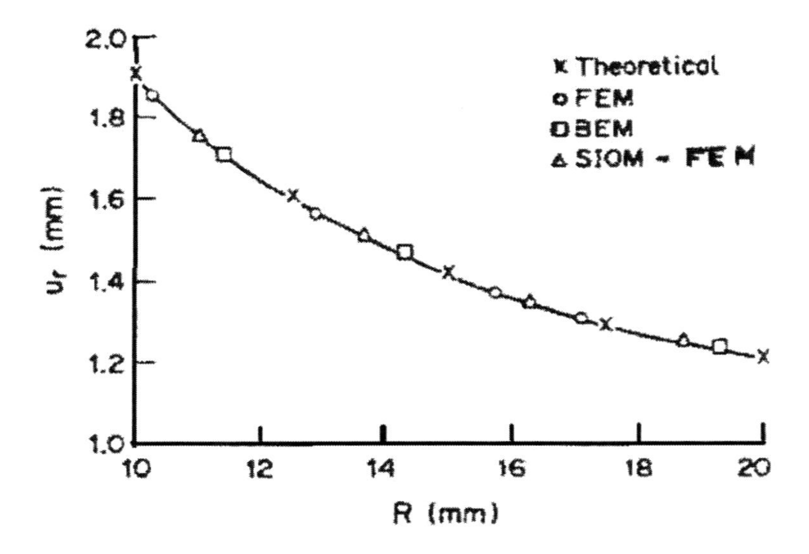

Figure 6.5.2. Relation between the radial displacements u_r and the radius R of the thick cylinder.

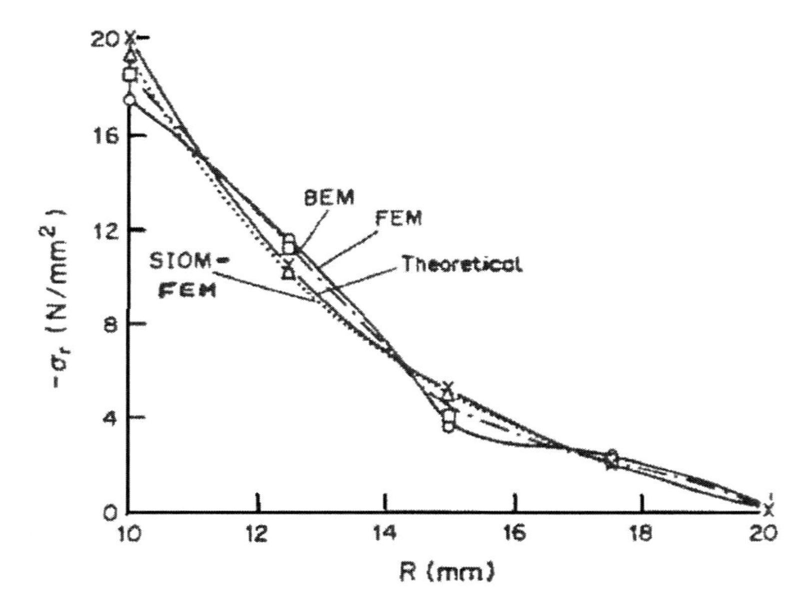

Figure 6.5.3. Relation between the radial stresses σ_r , and the radius R of the thick cylinder.

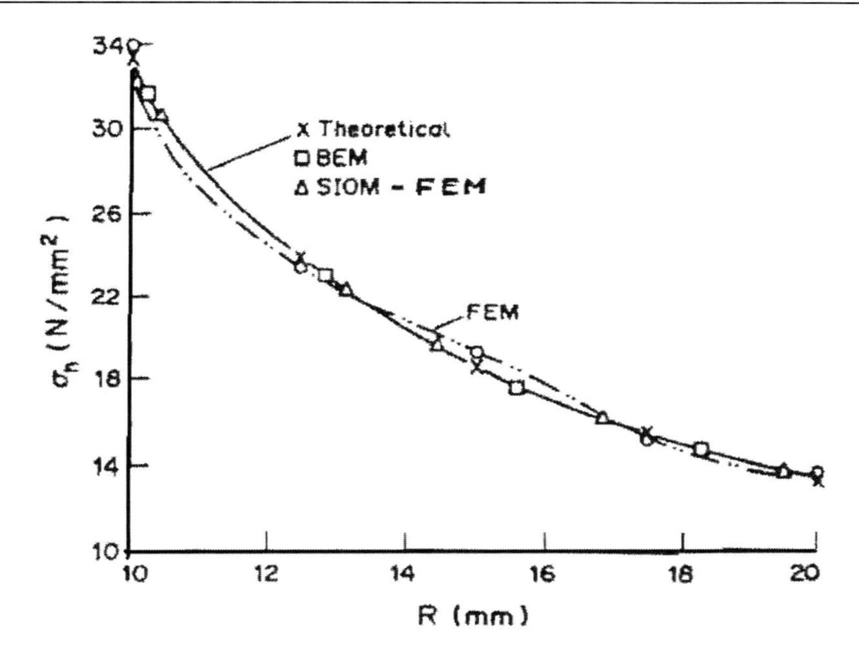

Figure 6.5.4. Relation between the hoop stresses σ_h and the radius R of the thick cylinder.

Also, Figure 6.5.3 shows the relation between the radial stresses σ_r and the radius R of the thick cylinder, by using theoretical methods and the numerical methods of the SIOM, FEM and BEM. Figure 6.5.4 shows further the relation between the hoop stresses σ_h and the radius R of the cylinder by using the theoretical method and the numerical methods of the SIOM, FEM and BEM.

As can be seen from Figs. 6.5.2 - 6.5.4, the computations of the SIOM and the theoretical results coincide very closely.

6.6. Three-Dimensional Application to the Determination of the Fracture Behaviour near the Edge of a Rectangular Crack under Tensile Pressure in an Isotropic Solid

As a second application of the three-dimensional elastic stress analysis, we will determine the fracture behaviour near the edge of a rectangular crack under tensile pressure p in an isotropic solid (Figure 6.6.1). This is a very specific problem of three-dimensional elasticity and has been previously solved by J. Weaver [42] by using another numerical method.

In the present section, the numerical results obtained with the Singular Integral Operators Method [45] will be referred to those pertinent to Weaver's numerical method.

Hence, let us determine the stress intensity factor K_1 near the edge of the crack given by the following formula [52]:

$$K_1 / \sqrt{d\pi} = \frac{u_3 G \sqrt{2}}{2r^{1/2}(1-v)} \tag{6.6.1}$$

where d is half side of the crack (see Figure 6.6.1), G denotes the shear modulus, v stands for Poisson's ratio, r represents the distance between the field point Y and the second order pole X (load point) and u_3 is the displacement in the z-direction (Figure 6.6.1).

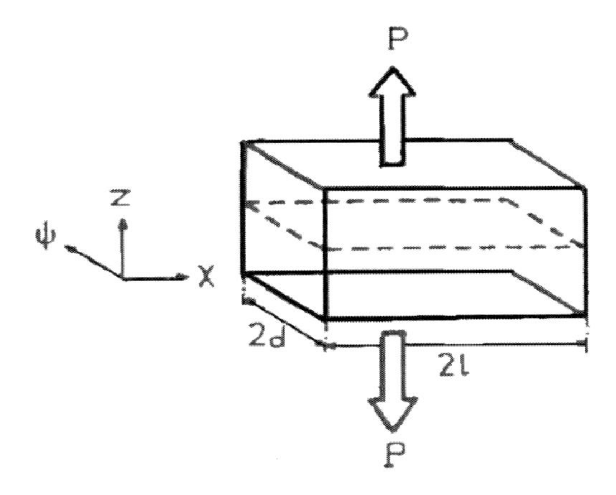

Figure 6.6.1. A rectangular crack under tensile pressure p in an isotropic solid.

If Δu_3 denotes the unknown dislocation density along the crack surface $S_{cr}^{\pm}$, then:

$$\Delta u_3 = 2u_3 \tag{6.6.2}$$

is valid.

By using (6.6.2), then (6.6.1) takes the form:

$$K_1 / \sqrt{d\pi} = \frac{\Delta u_3 G \sqrt{2}}{4r^{1/2}(1-v)} \tag{6.6.3}$$

On the other hand, the stress field σ_{i3} on the crack surface $S_{cr}^{\pm}$ for a crack of arbitrary shape, is valid as [42], [45]:

$$\sigma_{i3} = \frac{E(1-2v)}{16\pi(1-v^2)} \int_{S_{cr}^{\pm}} \left(\delta_{bi} r_{,c} - \delta_{ic} r_{,b} + \frac{3r_{,i} r_{,b} r_{,c}}{1-2v} \right) \frac{\Delta u_{b,c} \, \mathrm{d}S}{r^2}, \quad i = 1,2 \tag{6.6.4}$$

in which E is Young's modulus, $r_{,i} = \vartheta r / \vartheta x_i$, and δ denotes Kronecker's function and:

$$\sigma_{33} = \frac{E}{8\pi(1-v^2)} \int_{S_{cr}^{+}} \frac{r_{,i} \Delta u_{3,i} \, \mathrm{d}S}{r^2} \tag{6.6.5}$$

Moreover, for the present problem of tensile pressure p, it follows that:

$$p = -\sigma_{33} \tag{6.6.6}$$

Combining the above formulas, it is possible to compute the values of $K_1 / p\sqrt{\pi d}$, by using the Gauss-Legendre numerical integration rule.

Thus, Figure 6.6.2 shows the relation between the stress intensity factor $K_1 / p\sqrt{\pi d}$ and Y/d, for l/d = 1. Furthermore, Figs. 6.6.3 and 6.6.4 show the relation between the stress intensity factor $K_1 / p\sqrt{\pi d}$ and Y/d, for l/d = 3 and l/d = 5,7,9, respectively.

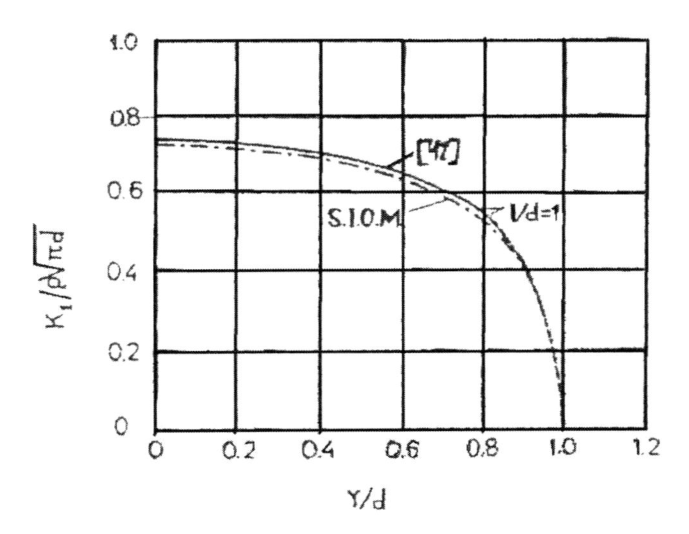

Figure 6.6.2. Relation between the stress intensity factor $K_1 / p\sqrt{\pi d}$ and Y/d for *l/d =1.*

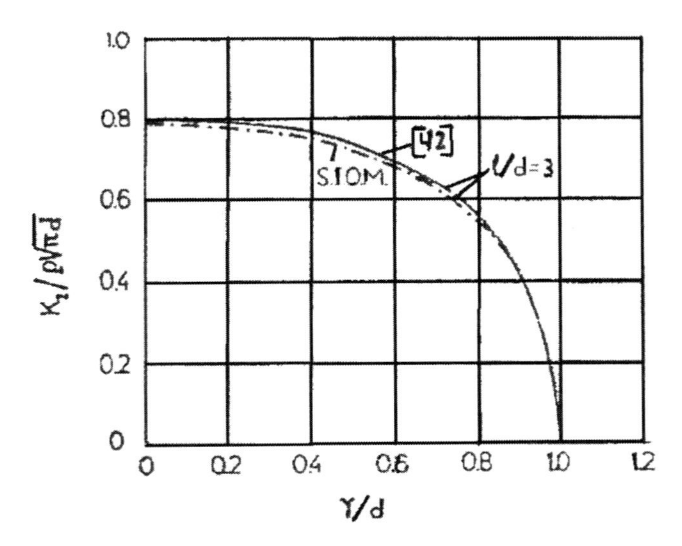

Figure 6.6.3. Relation between the stress intensity factor $K_1 / p\sqrt{\pi d}$ and Y/d for *l/d =3.*

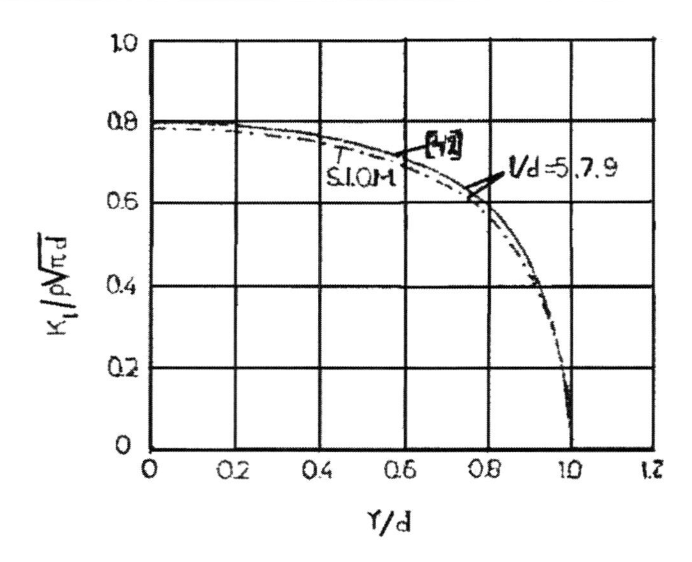

Figure 6.6.4. Relation between the stress intensity factor $K_1 / p\sqrt{\pi d}$ and Y/d for l/d =5, 7, 9.

Finally, Figure 6.6.5 shows the relation between the value of $K_1 / p\sqrt{\pi d}$ and X/d, for l/d = 1, l/d = 3, l/d = 5, l/d = 7 and l/d = 9.

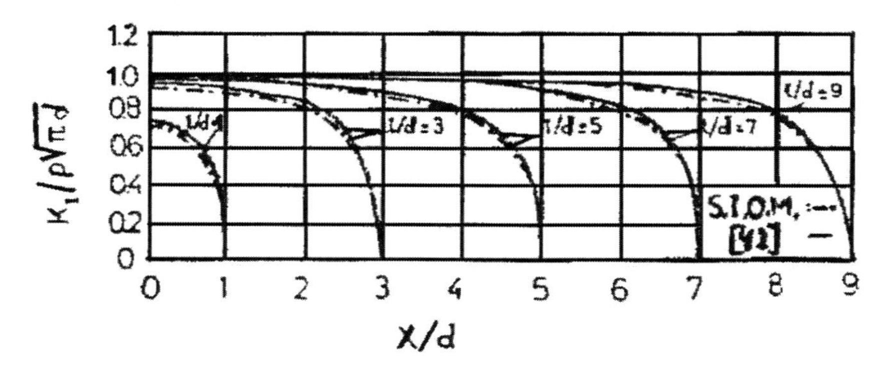

Figure 6.6.5. Relation between the stress intensityy factor $K_1 / p\sqrt{\pi d}$ and X/d for l/d =1, 3, 5, 7, 9.

As can be easily seen from Figs. 6.6.2, 6.6.3, 6.6.4 and 6.6.5, the computations of the Singular Integral Operators Method (S.I.O.M.) and the results of Weaver method [42] coincide very well.

6.7. Linear Theory of Viscoelasticity of Isotropic Solids

The multidimensional singular integral equations are further applied to the theory of linear viscoelasticity.

Some classical methods investigating problems of linear viscoelasticity have been proposed by G. A. C. Graham [53], T. L. Cost [54], W. T. Reed [55], E. H. Lee [56], R. A. Schapery [57] and T. C. Ting [58]. On the other hand, F. J. Rizzo and F. J. Shippy [59] have studied problems of linear viscoelasticity by using the Boundary Integral Equation Method (B.I.E.M.).

In the present section, linear viscoelasticity of isotropic solids will be reduced to the solution of multidimensional singular integral equations. [44]

Hence, consider a linear isotropic viscoelastic solid with the following stress field:

$$\sigma'_{ij} = 2\int_0^t V_1(F,t-\tau)\frac{\vartheta\varepsilon'_{ij}}{\vartheta\tau}(F,\tau)\,\mathrm{d}\,\tau + V_1(F,t)\varepsilon'_{ij}(F,0^+) \tag{6.7.1}$$

$$\sigma_{ii} = 3\int_0^t V_2(F,t-\tau)\frac{\vartheta\varepsilon_{ii}}{\vartheta\tau}(F,\tau)\,\mathrm{d}\,\tau + V_2(F,t)\varepsilon_{ii}(F,0^+) \tag{6.7.2}$$

with:

$$\varepsilon'_{ij}(F,0^+) = \lim_{t\to 0}\varepsilon_{ij}(F,t), \quad \varepsilon_{ii}(F,0^+) = \lim_{t\to 0}\varepsilon_{ii}(F,t) \tag{6.7.3}$$

in which σ'_{ij} and ε'_{ij} are, respectively, deviatoric components of the stress and strain field, V_1 and V_2 are relaxation functions in shear and isotropic compression, respectively, t is the time and F denotes a point in the infinite elastic space where the viscoelastic solid belongs.

By taking the Laplace transform of eqs. (6.7.1) and (6.7.2), one obtains:

$$\sigma'^*_{ij} = 2sV_1^*(s)\varepsilon'^*_{ij} \tag{6.7.4}$$

$$\sigma^*_{ii} = 3sV_2^*(s)\varepsilon^*_{ii} \tag{6.7.5}$$

Beyond the above, the Laplace transformed equation of equilibrium is as following:

$$\sigma^*_{ij,j} = 0 \tag{6.7.6}$$

Let us use Hooke's law for a linear isotropic, viscoelastic solid:

$$\sigma^*_{ij}(F) = \lambda^* u^*_{n,n}(F)\delta_{ij} + \mu^*[u^*_{i,j}(F) + u^*_{j,i}(F)] \tag{6.7.7}$$

where u^*_i denotes the boundary displacements, λ^* and μ^* the additional Lamé elastic constants and δ_{ij} is the delta function of Kronecker.

By applying the Betti-Rayleigh theorem the boundary displacement yields:

$$u^*_i(F) = -\int_L u^*_j(H)T_{ij}(H,F)\,\mathrm{d}\,H + \int_L t^*_j(H)U_{ij}(H,F)\,\mathrm{d}\,H \tag{6.7.8}$$

in which t_j^* are the boundary tractions on the contour L and H, F are points on the union of the contour L.

In (6.7.8) the fundamental solutions for the displacements and tractions are valid as following:

$$U_{ij} = \frac{\lambda^* + 3\mu^*}{8\pi\mu^*(\lambda^* + 2\mu^*)}\left(\frac{1}{r}\right)\left[\delta_{ij} + \left(\frac{\lambda^* + \mu^*}{\lambda^* + 3\mu^*}\right) r_{,i} r_{,j}\right] \tag{6.7.9}$$

$$T_{ij} = -\frac{\mu^*}{4\pi(2\mu^* + \lambda^*)}\left(\frac{1}{r^2}\right)\left[\frac{\vartheta r}{\vartheta n}\left(\delta_{ij} + \frac{3(\mu^* + \lambda^*)}{\mu^*} r_{,i} r_{,j}\right) - n_j r_{,i} + n_i r_{,j}\right] \tag{6.7.10}$$

By combining eqs. (6.7.4), (6.7.5) and (6.7.7) we obtain:

$$\lambda^* = -2/3 s V_1^*(s) + s V_2^*(s) \tag{6.7.11}$$
$$\mu^* = s V_1^*(s)$$

Finally, for small strain, the transformed components of strain and displacement are related by the formula:

$$2\varepsilon_{ij}^* = u_{ij,i}^* + u_{ji,i}^* \tag{6.7.12}$$

where u_i^* is given by (6.7.8).

Therefore, the components of the stress field for a linear, isotropic, viscoelastic solid can be evaluated by using the numerical technique of Chapter 5.

6.8. APPLICATION OF LINEAR VISCOELASTICITY

As an application of the previous theory, let us consider a thick, hollow circular cylinder of viscoelastic material restrained by an enclosing thin elastic ring and subjected to a uniform pressure p, applied as a step in time at $t = 0$ (see: Figure 6.8.1).

The same problem has been previously solved by T.C. Ting [58] by using an exact analytical solution and by F. J. Rizzo and D. J. Shippy [59] by using the Boundary Integral Equation Method (B.I.E.M.). A comparison will be made between the new Singular Integral Operators Method (S.I.O.M.) introduced in the present report, the theoretical solution [58] and the B.I.E.M. [59].

The geometrical sizes of the cylinder are the following: Ratio of inner to outer radii of the cylinder is $R_2/R_1 = 0.3$ (see: Figure 6.8.1).

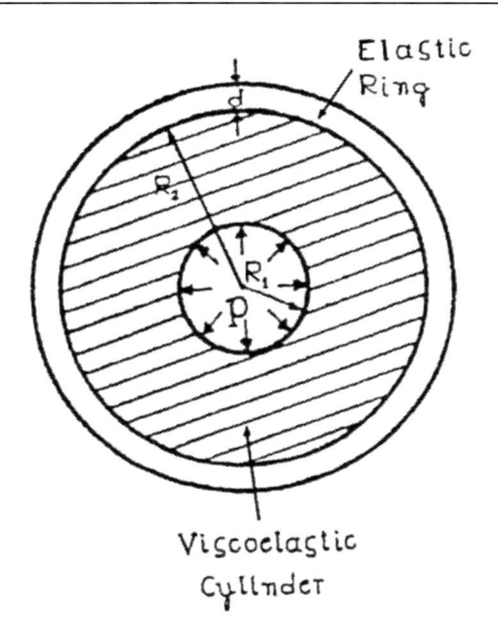

Figure 6.8.1. A thick, hollow circular cylinder of viscoelastic solid, restrained by an enclosing thin elastic ring and subjected to a uniform pressure p.

Beyond the above, we consider that the viscoelastic solid behaves elastically in bulk and as a standard linear solid in shear, so that the relaxation function V_1 and V_2 are given by the formulas:

$$V_1(t) = V\left[g + (1-g)e^{-\lambda t}\right] \tag{6.8.1}$$

$$V_2(t) = N \tag{6.8.2}$$

where g, λ, V and N are constants and we assume that $g = 0.5$ and $V = 0.6\,N$.

Therefore, the transformed relaxation functions V_1^* and V_2^* are given by the following relations:

$$V_1^* = (V/s)(s + g\lambda)/(s + \lambda) \tag{6.8.3}$$

$$V_2^* = N/s \tag{6.8.4}$$

Moreover, the thin, elastic ring is characterized by the relation:

$$C = Ed/(1-v^2)R_2 \tag{6.8.5}$$

where E is the Young's modulus, v Poisson's ratio, and d its thickness (Figure 6.8.1). Also we assume the value of C/N to be unity: $C/N = 1$

On the other hand, Figures 6.8.2 and 6.8.3 show the radial σ_r/p and the transverse stresses σ_θ/p, respectively, as functions of time.

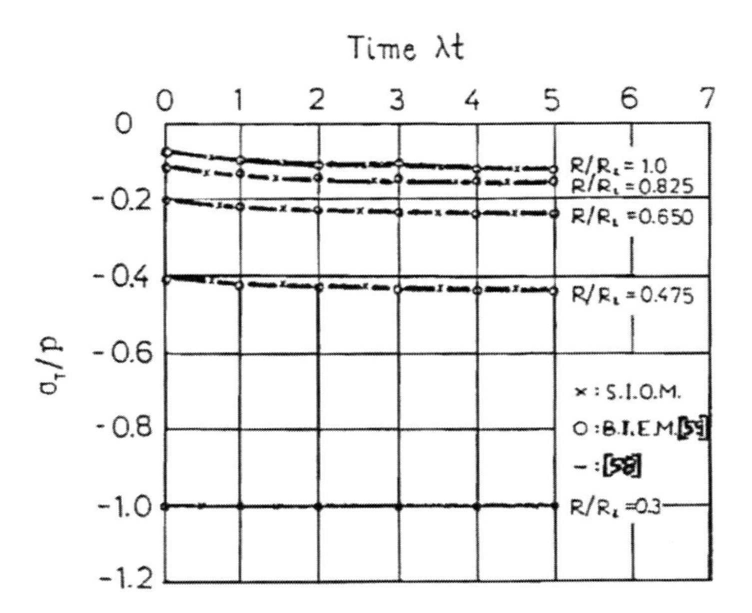

Figure 6.8.2. Radial stresses σ_r/p as functions of time for the cylinder of Figure 6.8.1.

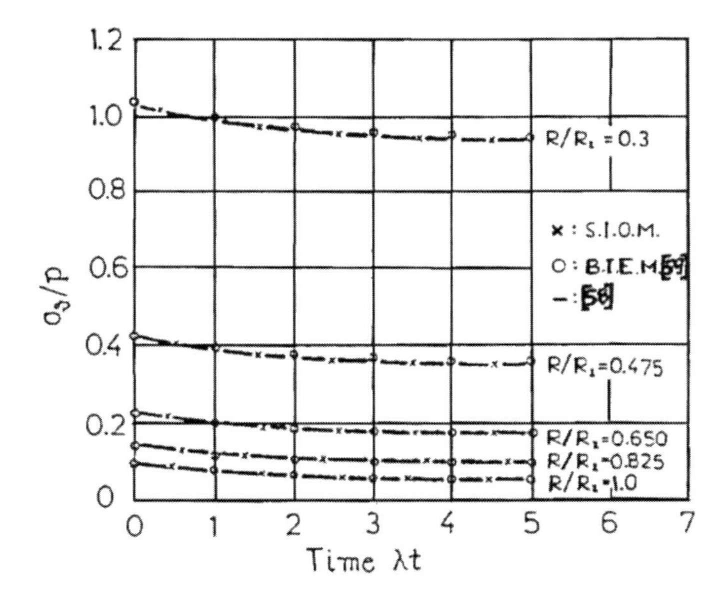

Figure 6.8.3. Transverse stresses σ_θ/p as functions of time for the cylinder of Figure 6.8.1.

Finally, as it is easily seen from Figures 6.8.2 and 6.8.3, the numerical results of the S.I.O.M. coincide very well with the theoretical results [58] and the corresponding numerical results of the B.I.E.M. [59]

REFERENCES

[1] V.D. Kupradze, Potential Methods in the Theory of Elasticity, Nauka, Moscow (1965).

[2] V.D. Kupradze, Three-dimensional Problems in the Mathematical Theory of Elasticity and Thermoelasticity, *Nauka,* Moscow (1976).

[3] M.H. Aliabadi, D.P. Rooke and D.J. Cartwright, An improved boundary element formulation for calculating stress intensity factors: Application to aerospace structures, *J. Str. Anal.* 22, 203-207 (1987).

[4] N. Altiero and D. Sikarskie, A boundary integral method applied to plates of arbitrary plane form, *Comp. Struct.* 9, 163-168 (1978).

[5] J. Balas, V. Sladek and J. Sladek, The boundary integral equation method for plates resting on a two parameter foundation, *ZAMM* 51, 574-580 (1984).

[6] P.K. Banerjee, The Boundary Element Methods in Engineering, McGraw-Hill, Berkshire (1994).

[7] J. Batista de Paiva, Boundary element formulation of building slabs, *Engng Anal. Bound. Elem.* 17, 105-110 (1996).

[8] G. Bezine, Boundary integral equations for plate flexure with arbitrary boundary conditions, *Mech. Res. Comm.* 5, 197-206 (1978).

[9] C.A. Brebbia, The Boundary Element Method for Engineers, Pentech Press, London (1978).

[10] C.A. Brebbia, The Boundary Element Method for Engineers, 2nd revised ed., Pentech Press, London (1980).

[11] C.A. Brebbia, Progresses in the Boundary Element Method, Pentech Press, London (1981).

[12] C.A. Brebbia, J.C.F. Telles and L.C. Wrobel, Boundary Element Techniques : Theory and Applications in Engineering, Springer-Verlag, Berlin (1984).

[13] H.D. Bui, Some remarks about the formulation of three-dimensional thermoelastoplastic problems of integral equations, *Int. J. Sol. Struct.* 14, 935-939 (1978).

[14] H.B. Chen, P. Lu, M.G. Huang and F.W. Williams, An effective method for finding values on and near boundaries in the elastic BEM, *Comp. Struct.* 69, 421-431 (1998).

[15] W.H. Chen and T.C. Chen, An efficient dual boundary element method technique for a two-dimensional fracture problem with multiple cracks, *Int. J. Numer. Meth. Engng* 38, 1739-1756 (1995).

[16] S. Christiansen and E. Hansen, A direct integral equation method for compounding the hoop stress at holes in plane, isotropic sheets, *J. Elasticity* 5, 1-14 (1975).

[17] T.A. Cruse, Application of the boundary-integral equation method to three-dimensional stress analysis, *Comp. Struct.* 3, 509-527 (1973).

[18] T.A. Cruse and W. Vanburen, Three-dimensional elastic stress analysis of a fracture specimen with an edge crack, *J. Fract. Mech.* 7, 1-15 (1971).

[19] M. Guiggiani, G. Krishnasamy, T.J. Rudolphi and F.J. Rizzo, A general algorithm for the numerical solution of hypersingular boundary integral equations, *ASME J. Appl. Mech.* 59, 604-614 (1992).

[20] G.A. Hartley and A. Abdel-Akher, Analysis of building frames, *ASCE J. Struct. Engng* 119, 468-483 (1993).

[21] G.A. Hartley, Development of plate bending elements for frame analysis, Engng Anal. *Bound. Elem.* 17, 93-104 (1996).

[22] M. Heinlein, S. Mukherjee and O. Richmond, A boundary element method analysis of temperature fields and stress during solidification, *Acta Mech.* 59, 58-81 (1986).

[23] U. Heise, The spectra of some integral operators for plane elastostatic boundary value problems, *J. Elasticity* 8, 47-49 (1978).

[24] J.C. Lachat, A further development of the boundary integral technique for elastostatics, Ph.D. thesis, Southampton University (1975).

[25] V. Mantic, A new formula for the C-matrix in the Somigliana identity, *J.Elasticity* 33, 191-201 (1993).

[26] D. Martin and M. Aliabadi, A BE hyper-singular formulation for contact problems using non-conforming discretization, *Comp. Struct.* 69, 557-565 (1998).

[27] M. Morjaria and S. Mukherjee, Numerical analysis of planar, time-dependent inelastic deformation of plates with cracks by the boundary element method, *Int. J. Sol. Struct.* 17, 127-143 (1981).

[28] S. Mukherjee, Corrected boundary integral equation in planar thermoelastoplasticity, *Int. J . Sol. Struct.* 13, 331-335 (1977).

[29] Y.X. Mukherjee, S. Mukherjee, X. Shi and A. Nagarajan, The boundary contour method for three-dimensional linear elasticity with a new quadratic boundary element, *Engng Anal. Bound. Elem.* 20, 35-44 (1997).

[30] Y.X. Mukherjee, K. Shah and S. Mukherjee, Thermoelastic fracture mechanics with regularized hypersingular boundary integral equations, *Engng Anal. Bound. Elem.* 23, 89-96 (1999).

[31] A.Nagarajan, S. Mukherjee and E. Lutz, The boundary contour method for three dimensional linear elasticity, *ASME J. Appl. Mech.* 63, 278-286 (1996).

[32] F. Paris and S.de León, Thin plates by the boundary element method by means of two Poisson equations, *Engng Anal. Bound. Elem.* 17, 111-122 (1996).

[33] N.N.V. Prasad, M.H. Aliabadi and D.P. Rooke, The dual boundary element method for thermoelastic crack problems, *Int. J. Fract.* 66, 255-272 (1994).

[34] N.N.V. Prasad, M.H. Aliabadi and D.P. Rooke, The dual boundary element method for transient thermoelastic crack problems, *Int. J. Solids Struct.* 33, 2695-2718 (1996).

[35] Y.F. Rashed and M.H. Aliabadi, Fundamental solutions for thick foundation plates, *Mech. Res. Commun.* 24, 331-340 (1997).

[36] Y.F. Rashed, M.H. Aliabadi and C.A. Brebbia, A boundary element formulation for a Reissner plate on a Pasternak foundation, *Comp. Struct.* 70, 515-532 (1999).

[37] D. Segond and A. Tafreshi, Stress analysis of three-dimensional contact problems using the boundary element method, *Engng Anal. Bound. Elem.* 22, 199-214 (1998).

[38] V. Sladek, J. Sladek and M. Tanaka, Evaluation of 1/r integrals in BEM formulations for 3-D problems using coordinate multitransformations, *Engng Anal. Bound. Elem.* 20, 229-244 (1997).

[39] M.A. Sutton, C.H. Liu, J.R. Dickerson and S.R. McNeill, The two-dimensional boundary integral equation method in elasticity with a consistent boundary formulation, *Engng Anal.* 3, 73-84 (1986).

[40] M. Tanaka, V. Sladek and J. Sladek, Regularization techniques applied to boundary element methods, *Appl. Mech. Rev.* 47, 457-499 (1994).

[41] A.N. Teong and D.L. Clements, A boundary integral equation method for the solution of a class of crack problems, *J. Elasticity* 17, 9-21 (1987).

[42] J. Weaver, Three-dimensional crack analysis, Int. J. Sol. Struct. 13, 321-330 (1977).

[43] N. Zabaras and S.M. Mukherjee, An analysis of solidification problems by the boundary element method, *Int. J. Num. Meth. Engng* 24, 1879-1900 (1987).

[44] E.G. Ladopoulos, On the numerical solution of the multidimensional singular integrals and integral equations used in the theory of linear viscoelasticity, *Int. J. Math. Math. Scien.* 11, 561-574 (1988).

[45] E.G. Ladopoulos, Cubature formulas for singular integral approximations used in three-dimensional elasticity, *Rev. Roum. Sci. Tech., Méc. Appl.* 34, 377-389 (1989).

[46] E.G. Ladopoulos, Singular integral operators method for three-dimensional elasto-plastic stress analysis, *Comp. Struct.* 38, 1-8 (1991).

[47] E.G. Ladopoulos, Singular integral operators method for two-dimensional elasto-plastic stress analysis, *Forsch. Ingen.* 57, 152-158 (1991).

[48] E.G. Ladopoulos, V.A. Zisis and D. Kravvaritis, Multidimensional singular integral equations in Lp applied to three-dimensional thermoelastoplastic stress analysis, *Comp. Struct.* 52, 781-788 (1994).

[49] E.G. Ladopoulos, Singular Integral Equations, Linear and Non-linear Theory and its Applications in Science and Engineering, Springer-Verlag, Berlin, New York (2000).

[50] N.I. Muskhelishvili, Singular Integral Equations, Groningen, Noordhoff, The Netherlands (1953).

[51] N.I. Muskhelishvili, Some Basic Problems of the Mathematical Theory of Elasticity, Groningen, Noordhoff, The Netherlands (1963).

[52] J.R. Rice, in Fracture (Ed.: H.Liebowitz), Vol.2, Chap.3, Academic Press, New York (1968).

[53] G.A.C. Graham, The correspondence principle of linear viscoelasticity theory for mixed boundary value problems involving time-dependent boundary regions, *Quart. Appl. Math.* 26, 167-174 (1968).

[54] T.L. Cost, Approximate Laplace transform inversions in viscoelastic stress analysis, *AIAA J .* 2, 2157-2166 (1964).

[55] W.T. Read, Stress analysis for compressible viscoelastic materials, *J. Appl. Phys.* 21, 671-674 (1950).

[56] E.H. Lee, Stress analysis in visco-elastic bodies, *Quart. Appl. Math.* 13, 183-190 (1965).

[57] R.A. Schapery, Approximate methods of transform inversion for viscoelastic stress analysis, *Proc. 4th U.S. Nat'l Congr. Appl. Math.* 2, 1075-1085 (1962).

[58] T.C. Ting, Remarks on linear viscoelastic stress analysis in cylinder problems, *Proc. 9th Midw. Mech. Conf.*, 263-275 (1965).

[59] F.J. Rizzo and D.J. Shippy, An application of the correspondence principle of linear viscoelasticity theory, *SIAM J. Appl. Math.* 21, 321-330 (1971).

Chapter 7

Computational Recipes for the Universal Equation of Elasticity and Thermo-Elasticity

7.1. Introduction

The mechanical foundations of the theory of thermo-elastic stress analysis for stationary frames began to be analysed in the early nineteenth century and were further developed during the twentieth century [1] - [23]. The modern resources for computer calculations have further brought within the realm of the practical the means of solving integral or differential equations and of making available a great number of useful solutions which were previously considered to be unmanageable. Past attempts to obtain solutions for several thermo-elasticity problems for stationary systems have led to a large variety of methods of solution, like singular integral equation methods, finite elements, boundary elements, integral transformation methods, etc.

Beyond the above, the previous methods dealing with the thermo-elastic stress behavior for stationary structures, shall be extended by the present study to the distribution of the thermo-elastic stress behavior for moving structures. Hence, a combination of the theories of elasticity and special relativity is presented, which leads to the investigation of the relativistic elastic stress analysis for moving (non-linear) frames. This results to the universal equation of elasticity.

During the past years special attention has been concentrated on the theoretical aspects of the special theory of relativity. Consequently, some classical monographs were written, dealing with the theoretical foundations and investigations of the special and the general theory of relativity [24] – [33].

On the other hand, by the present chapter the combination of the theories of thermoelasticity and special relativity leads to some applicable relativistic forms. This leads to the universal equation of thermo-elasticity. Thus, the stress field is investigated for moving structures under a big level of velocities. This level begins from very small velocities up to the approximation of the speed of light.

Hence, we shall show that for small velocities 50,000 km/h to 200,000 km/h, the absolute and the relative stress tensors are nearly the same. Also, for bigger velocities like c/3, c/2 or 3c/4 (c=speed of light) the relative stress tensor is much different from the absolute one, while for velocities near the speed of light the values of the relative stress tensor are very bigger than the corresponding values of the absolute stress tensor. For future aerospace applications the difference between the relative and the absolute stress tensors would be of very big interest.

Furthermore, a very important point which will be shown by the current research is that the relative stress tensor is not symmetrical, while, as it is well known, the absolute stress tensor is symmetrical. This difference is of big interest for future applications of aerospace structures.

Thus, the foundations of the new theory of relativistic thermo-elasticity for non-linear structures lead to a general theory, in which no restriction is made with regard to the relative motion. This general theory is further reduced to one class of relative motion, uniform in direction and velocity. Such relativistic stress analysis was defined and investigated by E.G. Ladopoulos [34]-[37].

Also, the big evolution of the jet engine and the high performance axial-flow compressor have considerably increased the possibilities of turbomachines applied in aerospace structures. The concern of very light weight in the aircraft propulsion application, and the desire to achieve the highest possible isentropic efficiency by minimizing parasitic losses led inevitably to axial-flow compressors with cantilever airfoils of high aspect ratio. Moreover, the turbojet engines were found to experience severe vibration of the rotor blades at part speed operation. The increasing evolution of aeroelasticity in aircraft turbomachines continues to be under active investigation, driven by the needs of aircraft powerplant and turbine designers.

The special principle of relativity is valid for all physical laws and according to this theory, all physical phenomena should have the same course of development in all systems of inertia, and observes installed in different systems of inertia should thus, as a result of their experiments, arrive at the establishment of the same laws of nature.

If this is true, then the notion of absolute space obviously loses its meaning, since any system of inertia with equally good reason can claim to be the absolute system of reference. The validity of the principle of relativity for all physical phenomena makes it simple to be applied for the distribution of the elastic stress behaviour for aerospace structures.

7.2. Relativistic Elastic Stress Analysis

Theorem 7.1.

Consider an infinitesimal face element ds with a directed normal, defined by a unit vector **n**, at definite point p in the three-space of a Lorenz system. The solid on either side of this face element experiences a force defined by:

$$\mathrm{d}\boldsymbol{\sigma}(\mathbf{n}) = \boldsymbol{\sigma}(\mathbf{n})\,\mathrm{d}s \tag{7.2.1}$$

where the components σ_i(n) of the force σ(n) are linear functions of the components n_k of n:

$$\sigma_i(\mathbf{n}) = \sigma_{ik} n_k, \quad i,k = 1,2,3 \tag{7.2.2}$$

in which σ_{ik} is the relative stress tensor of the moving system *S*, in contrast to the space part of the total energy-momentum tensor T_{ik}, referred as the absolute stress tensor σ_{ik}^{0} of the stationary system S^{0} (Figs 7.2.1 and 7.2.2).

Then, the connection between the absolute σ_{ik}^{0} and relative σ_{ik} stress tensors is given by the relation:

$$\sigma_{ik}^{0} = \sigma_{ik} + g_i u_k, \quad i,k = 1,2,3 \tag{7.2.3}$$

where g_i denotes the momentum density and u_k the components of the velocity u of the solid in the moving system S at the place and time considered.

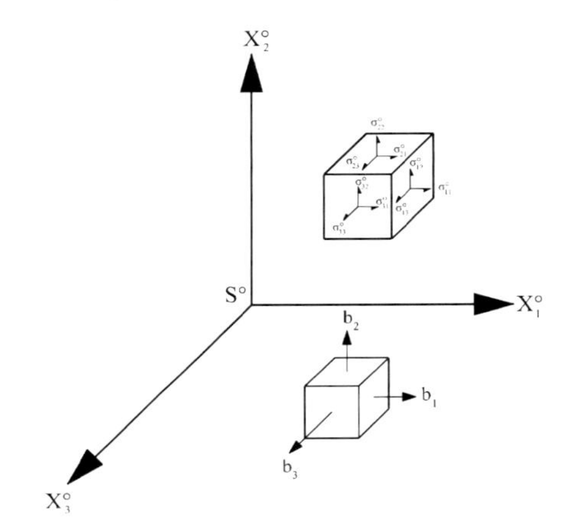

Figure 7.2.1. The state of the stress σ_{ik}^{0} in the stationary system S^{0}.

Proof.

The total elastic force F acting on the solid inside a closed surface s is valid as:

$$\int_s \boldsymbol{\sigma}(\mathbf{n})ds = \int_s \mathrm{d}\boldsymbol{\sigma}(\mathbf{n}) \tag{7.2.4}$$

where n is the inward normal to the surface element d*s*. By using Gauss's theorem and eqn (7.2.2), then the components F_i of this force are equal to:

$$F_i = \int_s \sigma_{ik} n_k \,\mathrm{d}s = -\int_V \frac{\partial \sigma_{ik}}{\partial x_k} \mathrm{d}V \tag{7.2.5}$$

in which *V* denotes a domain in three-dimensional space bounded by the closed surface *s*.

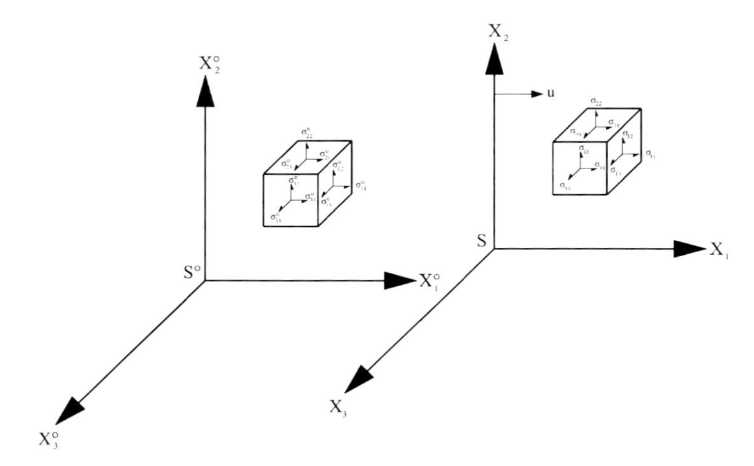

Figure 7.2.2. The state of the stress σ_{ik}^{0} in the stationary system S^{0} and σ_{ik} in the moving system with velocity u parallel to the x_1 - axis.

Hence, consider an elastic force density f such that:

$$F_i = \int_V f_i \mathrm{d}V \tag{7.2.6}$$

Then, by using (7.2.5) and (7.2.6) we obtain the following relation between the relative stress tensor σ_{ik} and the elastic force density f_i:

$$f_i = -\frac{\partial \sigma_{ik}}{\partial x_k} \tag{7.2.7}$$

Furthermore, the motion of an infinitesimal piece of matter with the volume δV is determined by the following equations of motion:

$$\frac{\mathrm{d}}{\mathrm{d}t}(g_i \delta V) = f_i \delta V = -\frac{\partial \sigma_{ik}}{\partial x_k}\delta V \tag{7.2.8}$$

in which g denotes the momentum density and d/dt the substantial time derivative.

Thus, eqn (7.2.8) may be written as following:

$$\frac{\mathrm{d}g_i}{\mathrm{d}t}\delta V + g_i \frac{\mathrm{d}}{\mathrm{d}t}\delta V = (\frac{\partial g_i}{\partial t} + \frac{\partial g_i}{\partial x_k}u_k)\delta V + g_i \delta V \frac{\partial u_k}{\partial x_k} \tag{7.2.9}$$

which is finally equal to:

$$(\frac{\partial g_i}{\partial t} + \frac{\partial_i (g_i u_k)}{\partial x_k}u_k)\delta V \tag{7.2.10}$$

Hence, by combining (7.2.8), (7.2.9) and (7.2.10) we obtain:

$$\frac{\partial g_i}{\partial t} + \frac{\partial}{\partial x_k}(g_i u_k + \sigma_{ik}) = 0 \tag{7.2.11}$$

Beyond the above, consider the law of conservation of energy and momentum:

$$\frac{\partial T_{ik}}{\partial x_k} = 0 \tag{7.2.12}$$

where T_{ik} denotes the total energy momentum tensor.

Therefore, eqn (7.2.12) for *i, k* = 1,2,3 can be also written as:

$$\frac{\partial \sigma_{ik}^{o}}{\partial x_k} + \frac{\partial g_i}{\partial t} = 0 \tag{7.2.13}$$

and thus, from (7.2.12) and (7.2.13) follows the required relation (7.2.3).

Theorem 2.2.

Consider the momentum density g defined by the relation (7.2.3) for the moving frame shown in Figure 7.2.1. Then, this density is given by the formula:

$$\mathbf{g} = m\mathbf{u} + \frac{(\mathbf{u}, \boldsymbol{\sigma})}{c^2} \tag{7.2.14}$$

where m is the total mass density:

$$m = E / c^2 \tag{7.2.15}$$

with E the total energy density, including the elastic energy, σ the relative stress tensor, u the velocity of the solid at the place and time considered and c the speed of light.

Proof.

The connection between the momentum density g and the energy flux D is given by the relation:

$$\mathbf{g} = \mathbf{D}/c^2 \tag{7.2.16}$$

where c denotes the speed of light.

Moreover, the total work per unit time done by the elastic forces on the solid, inside a closed surface s is valid as:

$$W = \int_s \left(\boldsymbol{\sigma}(\mathbf{n}), \mathbf{u}\right) \mathrm{d}s = \int_s \sigma_{ik} n_k u_i \, \mathrm{d}s = -\int_V \frac{\vartheta(u_i \sigma_{ik})}{\vartheta x_k} \mathrm{d}V \,, \quad i,k = 1,2,3 \tag{7.2.17}$$

in which n denotes the inward normal to the surface element ds and the integration in the last integral is extended over the interior V of the surface s.

Consequently, from eqn (7.2.17) we obtain the work done on an infinitesimal piece of matter of volume δW:

$$\delta W = -\frac{\partial(u_i \sigma_{ik})}{\partial x_k} \delta V \tag{7.2.18}$$

Furthermore, eqn (7.2.18) must be equal to the increase per unit time of the energy inside δV:

$$\frac{\mathrm{d}}{\mathrm{d}t}(E\delta V) = \delta W \tag{7.2.19}$$

where E denotes the total energy density, including the elastic energy. Eqn (7.2.9) can be further written as:

$$\frac{\mathrm{d}}{\mathrm{d}t}(E\delta V) = \left(\frac{\partial E}{\partial t} + \frac{\partial E}{\partial x_k} u_k \right) \delta V + E\delta V \frac{\partial u_k}{\partial x_k} \tag{7.2.20}$$

which is equal to:

$$\left[\frac{\partial E}{\partial t} + \frac{\partial}{\partial x_k}(Eu_k) \right] \delta V \tag{7.2.21}$$

and finally we obtain:

$$\frac{\partial E}{\partial t} + \frac{\partial}{\partial x_k}(Eu_k + u_i \sigma_{ik}) = 0 \tag{7.2.22}$$

Beyond the above, the general continuity equation for energy has the form:

$$div\mathbf{D} + \frac{\partial E}{\partial t} = 0 \tag{7.2.23}$$

Hence, by combining (7.2.22) and (7.2.23) follows:

$$\mathbf{D} = E\mathbf{u} + (\mathbf{u}, \boldsymbol{\sigma}) \tag{7.2.24}$$

in which $(\mathbf{u},\boldsymbol{\sigma})$ is the inner product with components $(\mathbf{u},\boldsymbol{\sigma})_k = u_i\sigma_{ik}$ and therefore, from (7.2.16) and (7.2.24) follows the required (7.2.14).

Corollary 2.1

The momentum density vector g does not in general have the same direction as the motion of the matter:

$$g_i u_k \neq g_k u_i \tag{7.2.25}$$

where u_i and u_k are the components of the velocity u of the solid at the place and time considered.

Proof.

The proof of the present Corollary follows directly from (7.2.14).

Corollary 2.2

The relative stress tensor σ_{ik} is not symmetrical.

Proof.

The law of conservation of angular momentum requires the absolute stress tensor σ_{ik}^o to be symmetrical:

$$\sigma_{ik}^o = \sigma_{ki}^o \tag{7.2.26}$$

Hence, from (7.2.3) , (7.2.14) and (7.2.26) follows:

$$\sigma_{ik} - \sigma_{ki} = -g_i u_k + g_k u_i = [-(\mathbf{u},\boldsymbol{\sigma})_i u_k + (\mathbf{u},\boldsymbol{\sigma})_k u_i]/c^2 \neq 0 \tag{7.2.27}$$

and therefore the relative stress tensor σ_{ik} is not symmetrical.

Corollary 2.3

In the stationary system S^O the following formulas are valid:

$$\sigma_{ik}^0 = \sigma_{ik} = \sigma_{ki} = \sigma_{ki}^0\ ,\ D_i^0 = g_i^0 = \sigma_{i4} = 0,\ \sigma_{44} = -E^0 \tag{7.2.28}$$

Proof.

As in the stationary system S^O (Figure 7.2.1) we have $\mathbf{u}^0 = 0$, and then directly from (7.2.3), (7.2.14) and (7.2.24) follows the required (7.2.28).

Theorem 2.3

The mechanical energy-momentum tensor T_{ik} is the sum of a kinetik part K_{ik} and a potential part P_{ik} :

$$T_{ik} = K_{ik} + P_{ik} \tag{7.2.29}$$

where:

$$K_{ik} = E^0 U_i U_k / c^2 = m^0 U_i U_k \tag{7.2.30}$$

in which U_i denotes the four-velocity of the matter and the potential part P_{ik} is determined completely by the relative stress tensor σ_{ik}, and will be called the stress four-tensor.

Proof.

The mechanical energy-momentum tensor T_{ik} satisfies the following relation:

$$T_{ik} U_k = -E^0 U_i \tag{7.2.31}$$

where Ui is the four-velocity of the solid and E^0 the energy density in the stationary system. Eqn (7.2.31) is easily proved in the stationary system by using (7.2.28) and putting $U_i^0 = (0,0,0,ic)$.

Furthermore, the following relations are valid:

$$U_i T_{ik} U_k / c^2 = U_i^0 T_{ik}^0 U_k^0 / c^2 = -T_{44}^0 = E^0(x_1) \tag{7.2.32}$$

and by introducing the tensor:

$$\Delta_{ik} = \delta_{ik} + U_i U_k / c^2 \tag{7.2.33}$$

which satisfies the relations:

$$U_i \Delta_{ik} = \Delta_{ik} U_k = 0 \tag{7.2.34}$$

then, we can form the following symmetrical tensor:

$$P_{ik} = \Delta_{in} T_{nm} \Delta_{mk} = P_{ki} \tag{7.2.35}$$

which is orthogonal to U_i :

$$U_i P_{ik} = P_{ik} U_k = 0 \tag{7.2.36}$$

Therefore, by combining (7.2.31), (7.2.32), (7.2.33) and (7.2.35) we obtain:

$$P_{ik} = T_{ik} - E^0 U_i U_k / c^2 \tag{7.2.37}$$

Moreover, in the stationary system the following relations are satisfied:

$$P_{ik}^0 = \sigma_{ik}^0 = \sigma_{ik}, \quad P_{i4}^0 = P_{4i}^0 = 0 \tag{7.2.38}$$

and (7.2.37) can be also written as (7.2.29).

Theorem 2.4.

The connection between the relative $\boldsymbol{\sigma}$ and the absolute $\boldsymbol{\sigma}^0$ stress tensors is given by the formula:

$$\boldsymbol{\sigma} = \boldsymbol{\sigma}^0 + \mathbf{u} \bullet (\boldsymbol{\sigma}^0, \mathbf{u}) \frac{(\gamma - 1)}{u^2} - (\boldsymbol{\sigma}^0, \mathbf{u}) \bullet \mathbf{u} \frac{(\gamma - 1)}{\gamma u^2} - (\mathbf{u} \bullet \mathbf{u})(\mathbf{u}, \boldsymbol{\sigma}^0, \mathbf{u}) \frac{(\gamma - 1)^2}{\gamma u^4} \tag{7.2.39}$$

where:

$$\gamma = 1/(1 - u^2 / c^2)^{1/2} \tag{7.2.40}$$

and u is the velocity of the solid at the place and time considered, c the speed of light and the notation $\mathbf{a} \bullet \mathbf{b}$ denotes the direct product of the spatial vectors a and b.

Proof.

In the stationary system the following tensor denotes a real symmetrical matrix:

$$P_{ik}^0 = \sigma_{ik}^0 = \sigma_{ik} \tag{7.2.41}$$

Hence, the normalized eigenvectors $\mathbf{d}^{(j)0}$, j = 1,2,3 satisfy the orthonormality and completeness relations:

$$(\mathbf{d}^{(j)0}, \mathbf{d}^{(l)0}) = \delta_{jl} \tag{7.2.42a}$$

and:

$$\mathbf{d}_i^{(j)0} \mathbf{d}_k^{(j)0} = \delta_{ik} \quad (j, l = 1,2,3) \tag{7.2.42b}$$

Furthermore, consider the principal stresses $p_{(j)}^0$ which are the three roots of the following equation:

$$\left|P_{ik}^{0} - \xi\delta_{ik}\right| = \left|\sigma_{ik}^{0} - \xi\delta_{ik}\right| = 0 \tag{7.2.43}$$

where ξ is the unknown.

The matrix P_{ik}^{0} may be further written in terms of the eigenvectors $\mathbf{d}^{(j)0}$ and eigenvalues $p_{(j)}^{0}$ as follows:

$$P_{ik}^{0} = p_{(j)}^{0} d_i^{(j)0} d_k^{(j)0} \tag{7.2.44}$$

Hence, in any system S one has:

$$P_{ik} = p_{(j)}^{0} d_i^{(j)} d_k^{(j)} \tag{7.2.45}$$

and:

$$\sigma_{ik} = P_{ik} - P_{i4} U_k / U_4 \tag{7.2.46}$$

By using (7.2.29), (7.2.30), (7.2.45) and (7.2.46), then we obtain:

$$P_{ik} = m^0 U_i U_k + p_{(j)}^{0} d_i^{(j)} d_k^{(j)} \tag{7.2.47}$$

and:

$$\sigma_{ik} = P_{ik} - P_{i4} U_k / U_4 = p_{(j)}^{0} d_i^{(j)} \left(d_k^{(j)} + i d_4^{(j)} u_k / c \right) \tag{7.2.48}$$

Therefore, by using the notation $\mathbf{a} \bullet \mathbf{b}$ for the direct product of the vectors a and b, then (7.2.48) can be further written as:

$$\boldsymbol{\sigma} = p_{(j)}^{0} \left[(\mathbf{d}^{(j)} \bullet \mathbf{d}^{(j)}) + \frac{i}{c} d_4^{(j)} (\mathbf{d}^{(j)} \bullet \mathbf{u}) \right], \quad j = 1,2,3 \tag{7.2.49}$$

where the vectors $\mathbf{d}^{(j)}$ satisfy the tensor relations:

$$d_i^{(j)} d_i^{(l)} = \delta_{jl} \tag{7.2.50}$$

and:

$$d_i^{(j)} d_k^{(j)} = \Delta_{ik} \tag{7.2.51}$$

where Δ_{ik} is given by (7.2.33).

Furthermore, if the spatial axes of the stationary and the moving systems have the same orientation, then we have:

$$\mathbf{d}^{(j)} = \mathbf{d}^{(j)0} + \mathbf{u}(\mathbf{u}, \mathbf{d}^{(j)0})(\gamma - 1)/u^2 \qquad (7.2.52)$$

$$d_4^{(j)} = i(\mathbf{u}, \mathbf{d}^{(j)0})\gamma/c$$

For the special case $i = k = 4$, we obtain from (7.2.47) and (7.2.52):

$$d = -P_{44} = -m^0 U_4^2 - p_{(j)}^0 (\mathbf{u}, \mathbf{d}^{(j)0})^2 \gamma^2/c^2 \qquad (7.2.53)$$

Moreover, in the stationary system, (7.2.49) reduces to:

$$\boldsymbol{\sigma}^0 = p_{(j)}^0 \left(\mathbf{d}^{(j)0} \bullet \mathbf{d}^{(j)0}\right) \qquad (7.2.54)$$

Hence, from eqn. (7.2.53) we obtain the following transformation law for the energy density:

$$d = \frac{d^0 + (\mathbf{u}, \boldsymbol{\sigma}^0, \mathbf{u}/c^2)}{1 - u^2/c^2} \qquad (7.2.55)$$

$$(\mathbf{u}, \boldsymbol{\sigma}^0, \mathbf{u}) = u_i \sigma_{ik}^0 u_k \qquad (7.2.56)$$

and for the mass density:

$$m = \frac{m^0 + (\mathbf{u}, \boldsymbol{\sigma}^0, \mathbf{u}/c^2)}{1 - u^2/c^2} \qquad (7.2.57)$$

Therefore, from (7.2.47) and (7.2.52) for $k = 4$ we obtain the following relation for the momentum density g:

$$\mathbf{g} = \mathbf{u}\left[d^0 + (\mathbf{u}, \boldsymbol{\sigma}^0, \mathbf{u})(1 - \gamma^{-1})/u^2\right]\gamma^2/c^2 + (\boldsymbol{\sigma}^0, \mathbf{u})\gamma/c^2$$

$$(\boldsymbol{\sigma}^0, \mathbf{u})_i = \sigma_{ik}^0 u_k \qquad (7.2.58)$$

Finally, by substituting (7.2.52) in (7.2.48) we obtain the required relation (7.2.39).

Corollary 2.4.

Consider the special case, where the motion of the matter at the point considered is parallel to the x_1-axis, i.e., u = $(u,0,0)$ (Figure 7.2.2). Then the connection between the absolute σ and the relative $\boldsymbol{\sigma}^0$ stress tensor is given by the Universal Equation of Elasticity:

$$\boldsymbol{\sigma} = \begin{bmatrix} \sigma_{11} & \sigma_{12} & \sigma_{13} \\ \sigma_{21} & \sigma_{22} & \sigma_{23} \\ \sigma_{31} & \sigma_{32} & \sigma_{33} \end{bmatrix} = \begin{bmatrix} \sigma_{11}^{0} & \gamma\sigma_{12}^{0} & \gamma\sigma_{13}^{0} \\ \frac{1}{\gamma}\sigma_{21}^{0} & \sigma_{22}^{0} & \sigma_{23}^{0} \\ \frac{1}{\gamma}\sigma_{31}^{0} & \sigma_{32}^{0} & \sigma_{33}^{0} \end{bmatrix} \tag{7.2.59}$$

where γ is given by (7.2.40).

Proof.

The required formula (7.2.59) follows immediately from (7.2.39) by putting u = $(u,0,0)$.

7.3 Relativistic Thermo-Elastic Stress Analysis

Theorem 3.1.

Consider a general system of continuously distributed ponderable or visible matter, inside which invisible heat conduction can take place, while the motion of the visible matter is described by the four-velocity U_i. Then the energy-momentum tensor of the general system is given by the relation:

$$T_{ik} = M_{ik} + H_{ik} \tag{7.3.1}$$

where M_{ik} is the mechanical part of the energy-momentum tensor and H_{ik} the heat part.

Moreover, the mechanical part M_{ik} is valid as following:

$$M_{ik} = d^0 U_i U_k / c^2 + P_{ik} \tag{7.3.2}$$

and the heat part:

$$H_{ik} = \left(U_i V_k + V_i U_k\right) / c^2 \tag{7.3.3}$$

where the four-vector V_i satisfies the relation:

$$V_i = -\Delta_{ik} T_{kj} U_j = -T_{ik} U_k - d^0 U_i \tag{7.3.4}$$

in which d^0 denote the normalized eigenvectors, Δ_{ik} is the tensor given by (7.2.33) and P_{ik} the potential part of the energy momentum tensor.

Proof.

The four-vector V_i is orthogonal to U_i:

$$U_i V_i = 0 \quad (7.3.5)$$

and therefore, we obtain:

$$V_i = (\mathbf{V}, i(\mathbf{V}, \mathbf{u})/c) \quad (7.3.6)$$

where $\mathbf{u}$ denotes the velocity of the matter.

Hence, in the stationary system, (7.3.6) reduces to:

$$V_i^0 = (\mathbf{V}^0, 0) \quad (7.3.7)$$

Furthermore, by replacing (7.2.33) into (7.2.35) and using (7.2.32) and (7.3.4), then we obtain instead of (7.2.37):

$$P_{ik} = T_{ik} - d^0 U_i U_k / c^2 - (U_i V_k + V_i U_k)/c^2 \quad (7.3.8)$$

Therefore, from (7.3.8) follows the required relation (7.3.1), instead of (7.2.29).

Theorem 3.2.

Consider the general system of continuously matter described in the previous theorem, inside which invisible heat conduction can take place, while the motion of the matter is described by the four-velocity U_i or by the velocity u_i.

Then, the connection between the energy-momentum tensor T_{ik} and the relative stress tensor σ_{ik} of the general system is given by the relation:

$$T_{ik} = g_i u_k + \sigma_{ik} + u_i \xi_k / c^2 \quad (7.3.9)$$

with:

$$\xi_k = U_4 (V_k - V_4 U_k / U_4)/ic \quad (7.3.10)$$

where V_k denotes the four-vector given by (7.3.4), g_i the momentum density and c the speed of light.

Proof.

The quantity ξ_k seems to be the most important part of ξ_{ik}:

$$\xi_{ik} = H_{ik} - H_{i4} U_k / U_4 = U_i \left(V_k - V_4 U_k / U_4 \right) / c^2 \tag{7.3.11}$$

Furthermore, ξ_k can be written by the following form by using (7.2.40) and (7.3.6):

$$\xi_k = (\boldsymbol{\xi}, 0) \tag{7.3.12}$$

with:

$$\boldsymbol{\xi} = \gamma \left[\mathbf{V} - \mathbf{u} (\mathbf{V}, \mathbf{u}) / c^2 \right] \tag{7.3.13}$$

Moreover, in the stationary system, $\boldsymbol{\xi}^0$ is equal to the heat current density $\mathbf{V}^0$:

$$\boldsymbol{\xi}^0 = \mathbf{V}^0 \tag{7.3.14}$$

By combining (7.3.10) and (7.3.11), then we obtain:

$$\xi_{ik} = U_i \, \xi_k / \gamma c^2 \tag{7.3.15}$$

Therefore, by using (7.2.48), (7.3.1), (7.3.2), (7.3.11) and (7.3.15), we have:

$$T_{ik} - T_{i4} U_k / U_4 = \sigma_{ik} + \xi_{ik} = \sigma_{ik} + U_i \, \xi_k / \gamma c^2 \tag{7.3.16}$$

which finally reduces to the required formula (7.3.9).

Lemma 3.1.

Consider the general system of continuously matter, inside which invisible heat conduction can take place. Then the momentum density g of this system is given by the Universal Equation of Thermo-Elasticity:

$$\mathbf{g} = m\mathbf{u} + \frac{(\mathbf{u}, \boldsymbol{\sigma})}{c^2} + \frac{\boldsymbol{\xi}}{c^2} \tag{7.3.17}$$

where u denotes the velocity of the matter at the place and time considered, σ the relative stress tensor, ξ is given by (7.3.13) and m is the total mass density given by (7.2.15).

Proof.

From (7.3.9), we obtain for the energy current density:

$$D_k = E u_k + u_i \sigma_{ik} + \xi_k \tag{7.3.18}$$

which can be further written as:

$$\mathbf{D} = E\mathbf{u} + (\mathbf{u}, \boldsymbol{\sigma}) + \boldsymbol{\xi} \tag{7.3.19}$$

Hence, from (3.19) by using the formula of the momentum density g:

$$\mathbf{g} = \mathbf{D}/c^2 \tag{7.3.20}$$

we obtain the required relation (7.3.17) which is a generalization of (7.2.14), for a general system with heat conduction.

7.4. Application of Relativistic Mechanics

A combination of the theories of thermo-elasticity and special relativity has been introduced, in order to investigate the thermo-elastic stress behavior for non-linear (moving) structures.

Table 7.4.1.

Velocity u	$\gamma = 1/\sqrt{1-u^2/c^2}$	Velocity u	$\gamma = 1/\sqrt{1-u^2/c^2}$
50,000 km/h	1.000000001	0.800c	1.666666667
100,000 km/h	1.000000004	0.900c	2.294157339
200,000 km/h	1.000000017	0.950c	3.202563076
500,000 km/h	1.000000107	0.990c	7.088812050
10E+06 km/h	1.000000429	0.999c	22.36627204
10E+07 km/h	1.000042870	0.9999c	70.71244596
10E+08 km/h	1.004314456	0.99999c	223.6073568
2x10E+8 km/h	1.017600788	0.999999c	707.1067812
c/3	1.060660172	0.9999999c	2236.067978
c/2	1.154700538	0.99999999c	7071.067812
2c/3	1.341640786	0.999999999c	22360.67978
3c/4	1.511857892	c	∞

Hence, the theory of relativity mainly known for its theoretical aspects, was directed to more applicable forms. Table 7.4.1 shows some characteristic values of the velocity for the moving structure in connection with the corresponding values of γ.

Thus, from the above Table follows that for small velocities 50,000 *km/h* to 200,000 *km/h*, $\gamma = 1$ and therefore the absolute and the relative stress tensors are nearly the same. On the other hand, for bigger velocities 50,000 *km/h* to $10^8\ km/h$ these two kinds of stress tensors begin to become different, while for velocities $c/3$, $c/2$ or $3c/4$ (c = speed of light), the variable γ takes bigger values and thus, the relative stress tensor is different than the absolute one.

Finally, for values of the velocity for the non-linear (moving) structure near the speed of light, the variable γ takes much bigger values, and at the end when the velocity becomes equal to the speed of light, then γ tends to the infinity.

REFERENCES

[1] Oden J.T., 'Finite Elements in Nonlinear Continua', McGraw Hill, New York (1972).
[2] Nowacki W., 'Thermoelasticity', Pergamon Press, Oxford (1962).
[3] Green A.E. and Zerna W., 'Theoretical Elasticity', Oxford Univ. Press, Oxford (1954).
[4] Sneddon I.N. and Lowengrub M., 'Crack Problems in the Classical Theory of Elasticity', J. Wiley, New York (1969).
[5] Love A.E.H., 'Mathematical Theory of Elasticity', Cambridge Univ. Press, Cambridge (1934).
[6] Liebowitz H., 'Fracture', Academic Press, New York (1968).
[7] Drucker D.C. and Gilman J.J., 'Fracture of Solids', J.Wiley, New York (1963).
[8] Muskhelishvili N.I., 'Some Basic Problems of the Mathematical Theory of Elasticity', Noordhoff, Groningen, Netherlands (1953).
[9] Sih G.C., 'Mechanics of Fracture', Noordhoff, Leyden, Netherlands (1975).
[10] Eringen A.C., 'Continuum Physics', Academic Press, New York (1972).
[11] Boley B.A. and Weiner J.H., 'Theory of Thermal Stresses', J.Wiley, New York (1960).
[12] Kupradze V.D., 'Three-dimensional Problems in the Mathematical Theory of Elasticity and Thermoelasticity', Nauka, Moscow (1976).
[13] Lekhnitskii S.G., 'Theory of Elasticity of an Anisotropic Elastic Body', Holden-Day, San Fransisco (1963).
[14] Wang C.C. and Truesdell C., 'Introduction to Rational Elasticity', Noordhoff, Groningen, Netherlands (1973).
[15] Truesdell C. and Noll W., 'The Non-linear Field Theories of Mechanics', Handbuch der Physic, Vol. III/3, Springer Verlag, Berlin (1965).
[16] Washizu K., 'Variational Methods in Elasticity and Plasticity', Pergamon Press, Oxford (1975).
[17] Ciarlet P.G., 'Topics in Mathematical Elasticity', North Holland, Amsterdam (1985).
[18] Lions J.L., 'Quelques Methodes de Resolution des Problemes aux Limites Non Lineaires, Dunod, Paris (1969).
[19] Duvant G. and Lions J.L., 'Les Inequations en Mecanique et en Physique', Dunod, Paris (1972).
[20] Fichera G., 'Boundary Value Problems of Elasticity with Unilateral Constraints', Handbuch der Physik, Vol. VIa/2, Springer Verlag, Berlin (1972).
[21] Germain P., 'Mecanique des Milieux Continus', Masson, Paris (1972).
[22] Gurtin M.E., 'Introduction to Continuum Mechanics', Academic Press, New York (1981).
[23] Gurtin M.E., 'The Linear Theory of Elasticity', Handbuch der Physik, Vol. IVa/2, Springer Verlag, Berlin (1972).
[24] Von Laue M., 'Die Relativitätstheorie', Vol. 1, Vieweg und Sohn, Braunschweig (1919).
[25] Möller C., 'The Theory of Relativity', Oxford University Press, Oxford (1972).
[26] Synge J.L., 'General Relativity', Clarendon Press, Oxford. (1972).
[27] Aharoni J., 'The Special Theory of Relativity', Clarendon Press, Oxford (1965).
[28] Adler R., 'Introduction to General Relativity', McGraw-Hill, New York (1965).

[29] Prokhovnik S.J., 'The Logic of Special Relativity', Cambridge University Press, London (1967).
[30] Gursey F., 'Relativity, Groups and Topology', Gordon and Breach, New York (1964).
[31] Gold T., 'Recent Developments in General Relativity', Pergamon Press, New York (1962).
[32] Rindler W., 'Special Relativity', Oliver and Boyd, Edinburgh (1966).
[33] Pirani F.A.E., 'Lectures on General Relativity', Vol.1, Prentice-Hall, New Jersey (1964).
[34] E.G. Ladopoulos, Relativistic elastic stress analysis for moving frames, *Rev. Roum. Sci. Tech., Méc. Appl.* 36, 195-209 (1991).
[35] E.G. Ladopoulos, Singular Integral Equations, Linear and Non-linear Theory and its Applications in Science and Engineering, Springer-Verlag, Berlin, New York (2000).
[36] E.G. Ladopoulos, Relativistic mechanics for airframes applied in aeronautical technologies, *Ad. Bound. Elem. Tech.* 10, 395-405 (2009).
[37] E.G. Ladopoulos, Relativistic elasticity and the universal equation of elasticity for next generation aircrafts and spacecrafts, *J. GJRE Aerosp. Scien.* 12, 1-10 (2012).

Chapter 8

ELASTICITY AND FRACTURE MECHANICS OF ANISOTROPIC SOLIDS BY MULTIDIMENSIONAL SINGULAR INTEGRAL EQUATIONS

8.1. INTRODUCTION

In modern high technology are used and find wide application many homogeneous solids like paper, textolite, plywood, delta wood, pine wood, metal systems, reinforced concrete and laminates, which are often anisotropic (or at least orthotropic from point to point). As anisotropic solids, may be further considered plates with artificially-made differences between the flexural rigidities for different directions. Plates strengthened by stiffening ribs and corrugated plates may be regarded as anisotropic materials, too. If a crack or hole is presented and is not associated with a plane or elastic symmetry, then the problem must be treated as one of general anisotropy.

During the past years, the stress fields in anisotropic solids have been widely studied, because numerous engineering materials under normal or loading conditions show different mechanical properties along certain preferred directions.

Among the scientists, who followed classical lines we shall mention the following: S. G. Lekhnitskii [1]-[5], G. N. Savin [6]-[10], M. O. Basheleishvili [11], [12], J. R. Willis [13], H. T. Rathod [14], S. Krenk [15], G. C. Sih and H. Liebowitz [16], G. C. Sih and M. K. Kassir [17] and G. C. Sih et al. [18].

Beyond the above, by using an integral transform method obtained by I.N. Sneddon [19], [20] the governing partial differential equation of anisotropic elasticity is solved, while G.E. Tupholme [21], D.D. Ang and M.L. Williams [22], O.L. Bowie and C.E. Freese [23] have studied some fracture mechanics problems of orthotropic media.

Singular integral equation methods for solving two- and three-dimensional problems of cracks and holes in anisotropic bodies have been investigated by F. J. Rizzo and D. J. Shippy [24], S. M. Vogel and F. J. Rizzo [25], M. D. Snyder and T. A. Cruse [26], [27], E. G. Ladopoulos [28]-[30], K. S. Parihar and S. Sowdamini [31], T. Mura [32], C. Ouyang and Mei-Zi Lu [33], R. P. Gilbert et al. [34], R. P. Gilbert and M. Schneider [35], R. P. Gilbert and R. Magnanini [36], U. Zastrow [37] - [39], L. J. Gray, D. Ghosh and T. Kaplan [40], and M. A. Sales and L. J. Gray [41].

Structural and engineer members are often loaded in ways that produce a three-dimensional stress state.

In general, although the loading may appear outwardly simple, a complex state of stress can exist inside the medium, particularly in the neighbourhood of the cracks and the holes, where the stresses can undergo sharp variations.

In the neighbourhood of cracks of irregular shapes or holes, the stress state is often triaxial in nature and the problem of predicting the surface of crack propagation is very difficult.

Thus, by the present chapter, all the methods which were applied to boundary value problems of homogeneous and piecewise homogeneous isotropic bodies, may be extended to anisotropic elastic bodies.

This requires, a further elaboration of the theory of the various kinds of fundamental solutions for systems of elliptic equations with discontinuous coefficients and also an extension of the theory of multidimensional singular integral equations to systems of equations having these fundamental solutions as their kernels. These problems, which present new difficulties, are of considerable interest and should be made the subject of future investigations.

Therefore, some parameters, such as intensity factors, incorporate stress levels, geometry and crack size, may be evaluated from the stress analysis of cracked structures. These parameters deduced from elastic fracture mechanics analysis can correlate crack growth rate in specimens and structures which see nominally - elastic loading. Elastic fracture mechanics analysis idealizes the physical crack or inclusion problem in three different ways : Firstly the crack plane is generally taken to be flat, secondly the crack is assumed to be sufficiently large that the local material microstructure can be modeled as a continuum and, thirdly inelastic crack tip effects such as plasticity are restricted to a sufficiently small volume that they can be neglected.

The validity of these assumptions must be verified on the basis of the actual behaviour of the cracked structure.

On the other hand, in some problems such as planar structures containing through cracks and holes, the stress and strain fields ahead of the cracks and the hole, may be approximated as two dimensions. In order to form the three-dimensional anisotropic problem, the Somigliana identity has been used which provides a formal representation for the displacement of an interior point in the body in terms of integrals involving boundary data.

By this theory it is possible to provide integral relations between boundary tractions and displacements, which in a well-posed boundary-value problem are not concurrently assigned over the entire surface.

8.2. BASIC RELATIONSHIPS FOR ANISOTROPIC ELASTIC STRESS ANALYSIS

We will express the stresses $(\sigma_x, \sigma_y, \sigma_z, \tau_{yz}, \tau_{zx}, \tau_{xy})$ in terms of strains $(\varepsilon_x, \varepsilon_y, \varepsilon_z, \gamma_{yz}, \gamma_{zx}, \gamma_{xy})$ through a set of constants C_{ij}, which are called the moduli of elasticity:

$$\begin{bmatrix} \sigma_x \\ \sigma_y \\ \sigma_z \\ \tau_{yz} \\ \tau_{zx} \\ \tau_{xy} \end{bmatrix} = \begin{bmatrix} C_{11} & C_{12} & C_{13} & C_{14} & C_{15} & C_{16} \\ C_{21} & C_{22} & C_{23} & C_{24} & C_{25} & C_{26} \\ C_{31} & C_{32} & C_{33} & C_{34} & C_{35} & C_{36} \\ C_{41} & C_{42} & C_{43} & C_{44} & C_{45} & C_{46} \\ C_{51} & C_{52} & C_{53} & C_{54} & C_{55} & C_{56} \\ C_{61} & C_{62} & C_{63} & C_{64} & C_{65} & C_{66} \end{bmatrix} \times \begin{bmatrix} \varepsilon_x \\ \varepsilon_y \\ \varepsilon_z \\ \gamma_{yz} \\ \gamma_{zx} \\ \gamma_{xy} \end{bmatrix} \tag{8.2.1}$$

On the other hand, in order to express the strains in terms of stresses, we will use another set of 36 constants a_{ij} $(i, j = 1,2,\ldots,6)$, known as the coefficients of deformation:

$$\begin{bmatrix} \varepsilon_x \\ \varepsilon_y \\ \varepsilon_z \\ \gamma_{yz} \\ \gamma_{zx} \\ \gamma_{xy} \end{bmatrix} = \begin{bmatrix} a_{11} & a_{12} & a_{13} & a_{14} & a_{15} & a_{16} \\ a_{21} & a_{22} & a_{23} & a_{24} & a_{25} & a_{26} \\ a_{31} & a_{32} & a_{33} & a_{34} & a_{35} & a_{36} \\ a_{41} & a_{42} & a_{43} & a_{44} & a_{45} & a_{46} \\ a_{51} & a_{52} & a_{53} & a_{54} & a_{55} & a_{56} \\ a_{61} & a_{62} & a_{63} & a_{64} & a_{65} & a_{66} \end{bmatrix} \times \begin{bmatrix} \sigma_x \\ \sigma_y \\ \sigma_z \\ \tau_{yz} \\ \tau_{zx} \\ \tau_{xy} \end{bmatrix} \tag{8.2.2}$$

Thus, by considering the case where the material is "transversely isotropic", which means, that it possesses an axis of elastic symmetry such that the material is isotropic in the planes normal to this axis, then the following formula is valid between the stresses and strains:

$$\begin{bmatrix} \varepsilon_x \\ \varepsilon_y \\ \varepsilon_z \\ \gamma_{yz} \\ \gamma_{zx} \\ \gamma_{xy} \end{bmatrix} = \begin{bmatrix} a_{11} & a_{12} & a_{13} & 0 & 0 & 0 \\ a_{12} & a_{11} & a_{13} & 0 & 0 & 0 \\ a_{13} & a_{13} & a_{33} & 0 & 0 & 0 \\ 0 & 0 & 0 & a_{44} & 0 & 0 \\ 0 & 0 & 0 & 0 & a_{44} & 0 \\ 0 & 0 & 0 & 0 & 0 & 2(a_{11} - a_{12}) \end{bmatrix} \times \begin{bmatrix} \sigma_x \\ \sigma_y \\ \sigma_z \\ \tau_{yz} \\ \tau_{zx} \\ \tau_{xy} \end{bmatrix} \tag{8.2.3}$$

where z is the direction of the elastic symmetry.

The coefficients of deformation in (8.2.3) are equal to: [1]

$$a_{11} = \frac{1}{E_1}, \quad a_{12} = -\frac{v_1}{E_1},$$
$$a_{33} = \frac{1}{E_2}, \quad a_{13} = -\frac{v_2}{E_2}, \tag{8.2.4}$$
$$a_{44} = \frac{1}{G_2}, \quad 2(a_{11} - a_{12}) = \frac{2(1+v_1)}{E_1} = \frac{1}{G_1}$$

in which E_1, G_1 and v_1 are the Young's modulus, shear modulus, and Poisson's ratio, respectively, in the plane of isotropy and E_2, G_2 and v_2 are the same quantities in the transverse direction.

By the same way in order to express the stress components in terms of strains for a "transversely isotropic" material we will obtain the following formula:

$$\begin{bmatrix} \sigma_x \\ \sigma_y \\ \sigma_z \\ \tau_{yz} \\ \tau_{zx} \\ \tau_{xy} \end{bmatrix} = \begin{bmatrix} C_{11} & C_{12} & C_{13} & 0 & 0 & 0 \\ C_{12} & C_{11} & C_{13} & 0 & 0 & 0 \\ C_{13} & C_{13} & C_{33} & 0 & 0 & 0 \\ 0 & 0 & 0 & C_{44} & 0 & 0 \\ 0 & 0 & 0 & 0 & C_{44} & 0 \\ 0 & 0 & 0 & 0 & 0 & \frac{1}{2}(C_{11} - C_{12}) \end{bmatrix} \times \begin{bmatrix} \varepsilon_x \\ \varepsilon_y \\ \varepsilon_z \\ \gamma_{yz} \\ \gamma_{zx} \\ \gamma_{xy} \end{bmatrix} \tag{8.2.5}$$

where the elastic moduli C_{ij} may be expressed by the following relations: [1],[2]

$$C_{11} = 2G_1\left(1 - v_2^2 \frac{E_1}{E_2}\right) \Big/ \left(1 - v_1 - 2v_2^2 \frac{E_1}{E_2}\right)$$
$$C_{12} = 2G_1\left(v_1 + v_2^2 \frac{E_1}{E_2}\right) \Big/ \left(1 - v_1 - 2v_2^2 \frac{E_1}{E_2}\right)$$
$$C_{13} = E_1 v_2 \Big/ \left(1 - v_1 - 2v_2^2 \frac{E_1}{E_2}\right) \tag{8.2.6}$$
$$C_{33} = E_2(1 - v_1) \Big/ \left(1 - v_1 - 2v_2^2 \frac{E_1}{E_2}\right)$$
$$C_{44} = G_2$$
$$\frac{1}{2}(C_{11} - C_{12}) = G_1$$

Furthermore, for isotropic materials it is valid $v_1 = v_2$, $E_1 = E_2$ and $G_1 = G_2$ and so the elastic moduli C_{ij} may be related to the Lamé coefficients λ and μ as:

$$C_{11} = C_{12} = \lambda + 2\mu$$
$$C_{12} = C_{13} = \lambda \tag{8.2.7}$$
$$C_{44} = \mu$$

The strain components in (8.2.5) are expressed by the formulas:

$$\varepsilon_x = \frac{\vartheta u_x}{\vartheta x}, \quad \varepsilon_y = \frac{\vartheta u_y}{\vartheta y}, \quad \varepsilon_z = \frac{\vartheta u_z}{\vartheta z}$$
$$\gamma_{yz} = \frac{\vartheta u_z}{\vartheta y} + \frac{\vartheta u_y}{\vartheta z} \tag{8.2.8}$$
$$\gamma_{zx} = \frac{\vartheta u_x}{\vartheta z} + \frac{\vartheta u_z}{\vartheta x}$$
$$\gamma_{xy} = \frac{\vartheta u_x}{\vartheta y} + \frac{\vartheta u_y}{\vartheta x}$$

in which u_x, u_y and u_z are the components of displacements in cartesian coordinates.

8.3. Fundamental Solutions of Anisotropic Stress Field Analysis

Consider a body in three-dimensional space, which has a bounding surface L. Then, according to Betti's reciprocal theorem and by considering absence of body forces, one has: [24], [29]

$$\int_L (u_i T_{ij} - t_i U_{ij})\,\mathrm{d}R + \int_\Gamma (U_i T_{ij} - t_i U_{ij})\,\mathrm{d}R = 0 \tag{8.3.1}$$

where dR is an element of surface area at R, which is a point on L. Furthermore, Γ is the boundary of the finite or infinite domain of space in coordinates x_1, x_2, x_3, in which exist the anisotropic elastic body.

This boundary Γ is a connected closed Lyapounov surface.

In (8.3.1) u_i and t_i are the displacement and traction components, $U_{ij}(x,y)$ the displacement at point x in response to a concentrated unit body force acting in the j coordinate direction at point y, and T_{ij} the suitable boundary tractions.

Betti's theorem (eq. (8.3.1)) results in Somigliana's identity [24]:

$$u_i(y) = \frac{1}{a}\int_L \left[t_j(x)U_{ij}(x,y) - u_j T_{ij}(x,y)\right]\cdot \mathrm{d}R \tag{8.3.2}$$

in which the point dependence is explicitly indicated and a is the magnitude of the force components. Beyond the above, the following two limiting formulas have to exist:

$$\lim_{\varepsilon \to 0} \int_{\Gamma} u_i T_{ij} \, \mathrm{d}R = a u_j(y) \tag{8.3.3}$$

$$\lim_{\varepsilon \to 0} \int_{\Gamma} t_i U_{ij} \, \mathrm{d}R = 0 \tag{8.3.4}$$

where ε is the radius of a sphere centre y, with the boundary Γ, and $u_j(y)$ the displacement at the origin corresponding to u_i and t_i on L.

In order to derive the formula of the fundamental solution, we adopt the method of decomposition into plane waves used in [42]. Thus, consider the function g, which is an arbitrary distribution and vanishes outside a finite sphere.

The next formula is a solution of the differential equation:

$$\Delta_y u(y) = g(y) \tag{8.3.5}$$

$$u(y) = \int_A g(x) \left(-\frac{1}{4\pi |x - y|} \right) \mathrm{d}x \tag{8.3.6}$$

in which Δy denotes the Laplacean with respect to y_i.

The following identity is easily seen to be:

$$\int_{|\zeta|=1} \left| (x_i - y_i) \zeta_i \right| \mathrm{d}R = 2\pi |x - y| \tag{8.3.7}$$

From (8.3.5) and (8.3.7) results following relation:

$$\Delta_y |x - y| = \frac{2}{|x - y|} \tag{8.3.7a}$$

Hence, from (8.3.5), (8.3.6) and (8.3.7a) one has:

$$g(y) = \frac{1}{16\pi^2} \Delta_y^2 \int_A \int_{|\zeta|=1} g(x) \left| (x_i - y_i) \zeta_i \right| \mathrm{d}R \, \mathrm{d}x \tag{8.3.8}$$

Also, let us consider the function $h(\zeta, p)$ which is given by the formula:

$$h(\zeta, p) = \int_{(x \cdot \zeta) = p} g(x) \, \mathrm{d}R \tag{8.3.9}$$

Beyond the above, the following is valid:

$$\int_A \int_{|\zeta|=1} g(x)\left|(x_i - y_i)\zeta_i\right| \mathrm{d}\,R\,\mathrm{d}\,x = \int_{|\zeta|=1} \mathrm{d}\,R \int_{-\infty}^{\infty} |p| \mathrm{d}\,p \int_{(x-y)\cdot\zeta=p} g(x)\,\mathrm{d}\,x$$

$$= \int_{|\zeta|=1} \mathrm{d}\,R \int_{-\infty}^{\infty} |p| h(\zeta, p + y\cdot\zeta)\,\mathrm{d}\,p \tag{8.3.10}$$

and:

$$\Delta_y \int_{-\infty}^{\infty} |p| h(\zeta, p + y\cdot\zeta)\,\mathrm{d}\,p$$

$$= \Delta_y \left[\int_{(y\cdot\zeta)}^{\infty} (p - y\cdot\zeta) h(\zeta, p) \right] \mathrm{d}\,p - \int_{-\infty}^{(y\cdot\zeta)} (p - y\cdot\zeta) h(\zeta, p)\,\mathrm{d}\,p = 2h(\zeta, y\cdot\zeta) \tag{8.3.11}$$

From (8.3.8), (8.3.10) and (8.3.11) one has:

$$g(y) = -\frac{1}{8\pi^2} \Delta_y \int_{|\zeta|=1} h(\zeta, y\cdot\zeta)\,\mathrm{d}\,R \tag{8.3.12}$$

On the other hand, by considering the case where:

$$g(y) = \delta(y) \tag{8.3.13}$$

then we obtain:

$$h(\zeta, y\cdot\zeta) = \delta(y\cdot\zeta) \tag{8.3.14}$$

From (8.3.12), (8.3.13) and (8.3.14) one obtains the expression for the three-dimensional delta function:

$$\delta(x - y) = -\frac{1}{8\pi^2} \Delta_y \int_{|\zeta|=1} \delta((x - y)\cdot\zeta)\,\mathrm{d}\,R \tag{8.3.15}$$

Therefore, from (8.3.15) we derive the fundamental solution for the displacements:

$$U_{ij}(x, y) = \frac{1}{8\pi^2} \Delta_y \int_{\zeta=1} W_{ij}(x, y, \zeta)\,\mathrm{d}\,R \tag{8.3.16}$$

where the function W_{ij} is given by:

$$W_{ij}(x,y,\zeta)=\begin{cases}W_{ij}(x,y,\zeta), & (x-y)\cdot\zeta>0\\ 0, & (x-y)\cdot\zeta\le 0\end{cases} \tag{8.3.17}$$

According to the Cauchy-Kowalewski theorem we have:

$$W_{ij}=P_{ij}(\zeta)(x_k-y_k)\zeta_k \tag{8.3.18}$$

Hence, from (8.3.16), (8.3.17) and (8.3.18) we have:

$$U_{ij}(x,y)=\frac{1}{8\pi^2}\Delta_y\int\limits_{\substack{|\zeta|=1\\(x-y)\cdot\zeta>0}}P_{ij}(\zeta)\cos\varphi\,\mathrm{d}R \tag{8.3.19}$$

By using (8.3.7), then (8.3.19) takes the simpler form:

$$U_{ij}(x,y)=\frac{1}{4\pi^2|x-y|}\int\limits_{\substack{|\zeta|=1\\(x-y)\cdot\zeta>0}}P_{ij}(\zeta)\cos\varphi\,\mathrm{d}R \tag{8.3.20}$$

where φ is the angle between the vectors x-y and ζ.

From (8.3.20) we derive a simpler form, if the part of the integration over the unit hemispherical shell of (8.3.20) involving the azimuthal angle, is carried out:

$$U_{ij}(x,y)=\frac{1}{8\pi^2|x-y|}\int\limits_{\substack{|\zeta|=1\\(x-y)\cdot\zeta>0}}P_{ij}(\zeta)\,\mathrm{d}s \tag{8.3.21}$$

in which ds denotes an element of arc length.

Thus, (8.3.21) gives the solution for the general case of three-dimensional elasticity.

Moreover, $P_{ij}(\zeta)$ in (8.3.21) is given by:

$$P_{ij}(\zeta)=\frac{1/2\,\varepsilon_{imn}\varepsilon_{jrs}Q_{mr}(\zeta)Q_{ns}(\zeta)}{\det Q} \tag{8.3.22}$$

and:

$$Q_{ik}(\zeta)=C_{ijkl}\zeta_j\zeta_l$$

where the constants C_{ijkl} are the elasticities, Q_{ij} is the characteristic matrix and the quantities ε_{imn} and det$Q$ are the alternating symbol and determinant of Q_{ij}, respectively. On the other hand, the suitable boundary tractions T_{ij} are given by the formula:

$$T_{im}(x,y) = C_{ijkl} U_{km}(x,y)_{,l} n_j \tag{8.3.22a}$$

where n_i are the components of the unit outward at the point x on L. Moreover, let us take a new point x^1 relative to the point x. Then for the vectors x, x^1 one has:

$$\mathbf{x}^1 = \mathbf{x} + \delta\mathbf{x} \tag{8.3.23}$$

By the same way, the new point ζ_1 relative to the point ζ is valid as:

$$\zeta^1 = \zeta + \delta\zeta \tag{8.3.24}$$

Hence, from (8.3.22a) we obtain in an analogous way, the displacement tensor:

$$U_{ij}(x^1, y) = \frac{1}{8\pi^2 |x^1 - y|} \oint_{|\zeta^1|=1} P_{ij}(\zeta^1)\,\mathrm{d}s \tag{8.3.25}$$

Therefore, from (8.3.23) and (8.3.24), eq. (8.3.25) takes the form:

$$U_{ij}(x^1, y) = \frac{1}{8\pi^2 |x + \delta x - y|} \oint_{|\zeta^1|=1} P_{ij}(\zeta + \delta\zeta)\,\mathrm{d}s \tag{8.3.26}$$

Furthermore, we introduce the expressions:

$$\lambda_i = \frac{x_i - y_i}{|x - y|} \tag{8.3.27}$$

and:

$$\lambda_i^1 = \frac{x_i^1 - y_i}{|x^1 - y|} = \frac{x_i + \delta x_i - y_i}{|x + \delta x - y|} \tag{8.3.28}$$

Then, it is easily shown that:

$$\delta\zeta_i = \frac{-\lambda_k^1 \zeta_k}{1 + \lambda_k \lambda_k^1} (\lambda_i^1 + \lambda_i) \tag{8.3.29}$$

The insertion of (8.3.27), (8.3.28), (8.3.29) into (8.3.26) results the displacements:

$$U_{ij,k} = -\frac{(x_k - y_k)}{8\pi^2 |x - y|^3} \oint_{|\zeta|=1} P_{ij}(\zeta)\,\mathrm{d}s - \frac{1}{8\pi^2 |x - y|^3} \int_{|\zeta|=1} \zeta_k \tag{8.3.30}$$

$$\times \frac{\left[(x_q - y_q)R_{jiq} + (x_r - y_r)R_{jir} + (x_s - y_s)R_{jis} + (x_t - y_t)R_{jit}\right]\mathrm{d}s}{\det Q} + \frac{1}{8\pi^2 |x-y|^3} \int_{|\zeta|=1} P_{ij}\zeta_k$$

$$\times \frac{\left[(x_1 - y_1)W_1 + (x_m - y_m)W_m + (x_n - y_n)W_n + (x_p - y_p)W_p + (x_r - y_r)W_r + (x_s - y_s)W_s\right]\mathrm{d}s}{\det Q}$$

Finally, (8.3.30) gives the solution for the general case of three-dimensional elasticity, while the boundary tractions P_{ij} are given by (8.3.22).

8.4. Two-Dimensional Application to the Determination of the Fracture Behaviour of a Circular Hole Subjected to a Uniform Boundary Shear Traction in an Infinite and Orthotropic Solid

As an application of the two-dimensional Singular Integral Operators Method, we will determine the fracture behaviour of a circular hole subjected to a uniform boundary shear traction p in an infinite and orthotropic solid (see Figure 8.4.1). The same problem has been previously solved by Lekhnitskii [1],[2] by using a theoretical method. A comparison will be made between the numerical results obtained by the Singular Integral Operators Method (S.I.O.M.) and the theoretical results in [2].

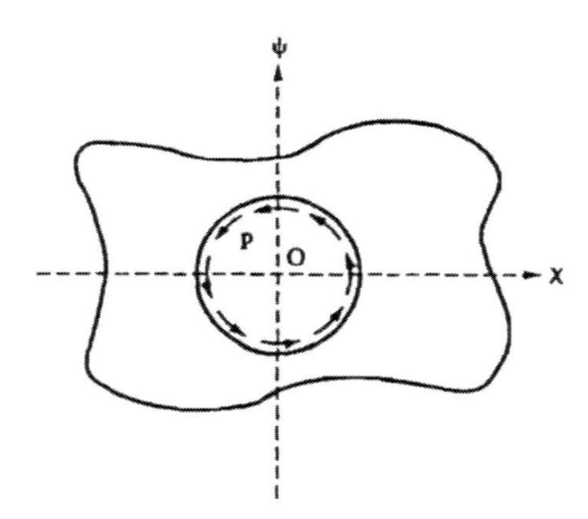

Figure 8.4.1. A circular hole subjected to a uniform boundary shear traction p in an infinite and orthotropic solid.

The solid under investigation was orthotropic with two mutually orthogonal axes of elastic symmetry in the plane. With respect to the principal material-axes of an orthotropic plate, the elastic constitutive expression relating the inplane stresses and strains is given as following: [1], [2]

$$\begin{bmatrix} \sigma_x \\ \sigma_y \\ \tau_{xy} \end{bmatrix} = \begin{bmatrix} C_{11} & C_{12} & 0 \\ C_{12} & C_{22} & 0 \\ 0 & 0 & C_{33} \end{bmatrix} \times \begin{bmatrix} \varepsilon_x \\ \varepsilon_y \\ \gamma_{xy} \end{bmatrix} \tag{8.4.1}$$

where the moduli of elasticity C_{ij} are equal to:

$$\begin{aligned} C_{11} &= E_1/(1-v_1v_2) \\ C_{12} &= v_2E_1/(1-v_1v_2) \\ C_{22} &= E_2/(1-v_1v_2) \\ C_{33} &= G \end{aligned} \tag{8.4.2}$$

in which E_1 and v_1 are the Young's modulus and Poisson's ratio in the plane of isotropy, while E_2 and v_2 are the same quantities in the transverse direction and G denotes the shear modulus.

Beyond the above, from the five elastic constants in (8.4.2), only four are independent, while the fifth constant may be evaluated from the equality:

$$v_2E_1 = v_1E_2 \tag{8.4.3}$$

The shear modulus G in eq. (8.4.2) can be computed by:

$$1/G = 4/E_{45} - (1-2v_1)/E_1 - 1/E_2 \tag{8.4.3a}$$

which in addition to E_1, E_2 and v_1, requires Young's modulus in the 45° direction.

The solid used was plywood with the following elastic constants: $E_1 = 1.2\times10^5$ *kg/cm*^{2}, $E_2 = 0.6\times10^5$ *kg/cm*^{2}, $G = 0.07\times10^5$ *kg/cm*^{2}, v_1=0.0712 and v_2=0.0356.

Beyond the above, according to Lekhnitski [2], for the case of a circular hole subjected to a uniform boundary shear traction p in an infinite and orthotropic solid, the stress field in the neighbourhood of the hole is valid as:

$$\sigma_\Theta = p\frac{E_\Theta}{2E_1}\sin 2\Theta\left[(1-k)(n-k-1)+(1+\mu_1^2)(1+\mu_2^2)\cos 2\Theta\right] \tag{8.4.4}$$

where μ_1 and μ_2 are the roots of the following equation:

$$\frac{\mu^4}{E_1} + \left(\frac{1}{G} - \frac{2v_1}{E_1}\right)\mu^2 + \frac{1}{E_2} = 0 \tag{8.4.5}$$

and k, n are valid as:

$$k = -\mu_1 \mu_2 = \left(E_1 / E_2\right)^{1/2} \tag{8.4.6}$$

$$n = -i(\mu_1 + \mu_2) = \left[2\left(\frac{E_1}{E_2} - \nu_1 \right) + \frac{E_1}{G} \right]^{1/2} \tag{8.4.7}$$

In (8.4.7) E_Θ denotes the Young's modulus in the direction tangent to the opening contour at angle Θ:

$$1/E_\Theta = \sin^4 \Theta / E_1 + \left(1/G - 2\nu_1 / E_1\right) \sin^2 \Theta \cos^2 \Theta + \cos^4 \Theta / E_2 \tag{8.4.8}$$

Hence, for the plywood used from (8.4.5), (8.4.6) and (8.4.7), one obtains: $\mu_1 = 4.11i$, $\mu_2 = 0.343i$, $k = 1.414$ and $n = 4.453$.

Table 8.4.1 shows the values of the stress tensor σ_Θ / p, by using the theoretical method (eq. 8.4.4) and the numerical by obtaining the results of the last two sections (S.I.O.M.).

As it is easily seen from Table 8.4.1, the theoretical and the numerical results (S.I.O.M.) compare very well.

Table 8.4.1. Stress field σ_Θ / p around the boundary of the hole of Figure 8.4.1

	σ_Θ / p	
Angle Θ from the x-axis	Theoretical	Numerical
(degrees)	[eq. (8.5.4)]	(S.I.O.M.)
0	0	0
5	-0.6064	-0.6093
10	-1.0077	-1.0563
15	-1.1563	-1.1602
20	-1.1185	-1.1236
25	-0.9743	-0.9856
30	-0.7775	-0.7703
35	-0.5565	-0.5500
40	-0.3241	-0.3195
45	-0.0844	-0.0832
50	0.1629	0.1701
55	0.4201	0.4267
60	0.6891	0.6923
65	0.9680	0.9665
70	1.2411	1.2406
75	1.4546	1.4583
80	1.4650	1.4592
85	1.0113	1.0104
90	0	0

Finally, Figure 8.4.2 shows the stress distribution along the opening edge. The stress is distributed nonuniformly and changes its values eight times.

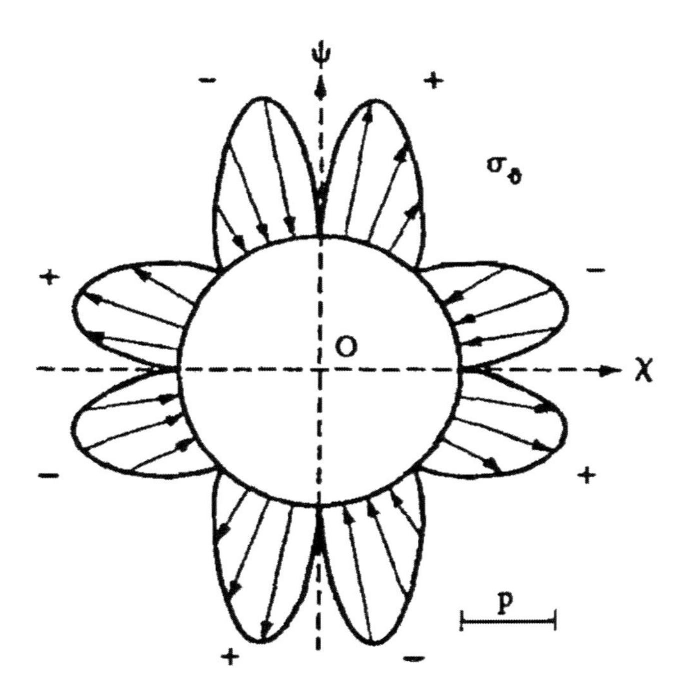

Figure 8.4.2. The stress distribution along the opening edge of Figure 8.4.1.

8.5. Two-Dimensional Application to the Determination of the Stress Field in a Circular Hole in the Centre of a Rectangular Orthotropic Plate, Subjected to Deformation by Tangential Forces

As a second application of the theory of anisotropic bodies, we will determine the stress distribution along the boundary of a circular hole in the centre of a rectangular orthotropic plate, subjected to deformation by tangential forces p which are uniformly distributed along its sides (see Figure 8.5.1). This problem has been theoretically solved by Lekhnitskii [1],[2] and a comparison will be made between the numerical (S.I.O.M.) and the theoretical results.

The solid used in the present application was plywood with the same elastic constants as in the previous section. According to Lekhnitskii [2], the stress field in the neighbourhood of the circular hole is valid as:

$$\sigma_\Theta = -p\frac{E_\Theta}{2E_1}(1+k+n)n\sin 2\Theta \tag{8.5.1}$$

where k, n and E_Θ are given by eqs. (8.4.6), (8.4.7) and (8.4.8), respectively.

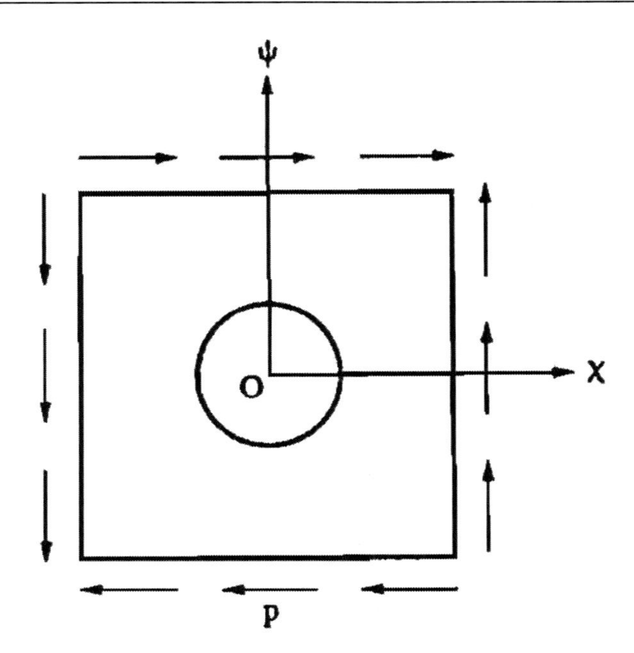

Figure 8.5.1. A circular hole in the centre of a rectangular orthotropic plate, subjected to deformation by tangential forces p which are uniformly distributed along its sides.

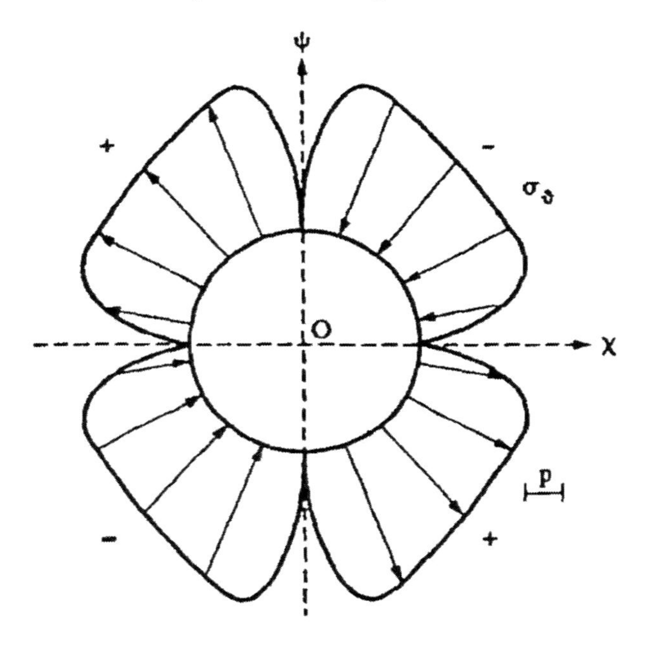

Figure 8.5.2. The stress distribution along the opening edge of Figure 8.5.1.

Table 8.5.1 shows the values of the stress field σ_{Θ}/p around the boundary of the circular hole by using the numerical method (S.I.O.M.) and the theoretical method (eq. (8.5.1)).

As it can be seen from Table 8.5.1, the theoretical and numerical results (S.I.O.M.) compare very well. Finally, Figure 8.5.2 shows the stress distribution along the opening edge, when the forces p are parallel to the principal directions of elasticity.

Table 8.5.1. Stress field σ_Θ/p around the boundary of the hole of Figure 8.5.1.

	σ_Θ/p	
Angle Θ from the x-axis (degrees)	Theoretical [eq. (8.5.1)]	Numerical (S.I.O.M.)
0	0	0
5	-1.2655	-1.2598
10	-2.1978	-2.1893
15	-2.7224	-2.7025
20	-2.9520	-2.9467
25	-3.0223	-3.0012
30	-3.0266	-3.0041
35	-3.0173	-3.0005
40	-3.0227	-3.0632
45	-3.0578	-3.0431
50	-3.1318	-3.1264
55	-3.2508	-3.2458
60	-3.4170	-3.4067
65	-3.6233	-3.6205
70	-3.8343	-3.8256
75	-3.9365	-3.9304
80	-3.6325	-3.6257
85	-2.3851	-2.3804
90	0	0

The singular integral operators method has been applied for the formulation of the anisotropic elastic stress components. By using the anisotropic theory it can be explained the mechanical behaviour of composite materials, which have increasingly wide application in engineering and have several possible fracture modes, such as fibre fracturing, crack bridging, matrix crazing and fibre-matrix debounding.

8.6. Crack Tip Analysis in an Anisotropic Solid

The strains are expressed in terms of stresses through a set of constants $a_{ij}\ (i, j = 1,2,3)$:

$$\begin{bmatrix} \varepsilon_x \\ \varepsilon_y \\ \gamma_{xy} \end{bmatrix} = \begin{bmatrix} a_{11} & a_{12} & a_{13} \\ a_{21} & a_{22} & a_{23} \\ a_{31} & a_{32} & a_{33} \end{bmatrix} \times \begin{bmatrix} \sigma_x \\ \sigma_y \\ \tau_{xy} \end{bmatrix} \tag{8.6.1}$$

On the other hand, according to Lekhnitski [1],[2] for the anisotropic solids it is valid the following characteristic equation:

$$a_{11}\mu^4 - 2a_{13}\mu^3 + (2a_{12} - a_{33})\mu^2 - 2a_{23}\mu + a_{22} = 0 \tag{8.6.2}$$

The four roots of (8.6.2) are never real and have the following form:

$$\mu_j = c_j \pm ib_j, \ j = 1, 2, \ldots \tag{8.6.3}$$

As it is well known , for the isotropic solids is valid $\mu = \pm 1$.

Let us further introduce a new system of coordinates z_j, so that:

$$z_j = x + \mu_j y, \ \ j = 1, 2 \tag{8.6.4}$$

Beyond the above, the fundamental solution for the displacements and tractions are given by the following formulas, for surface forces at point k acting in the l-direction: [24], [25], [28]

$$U_{lk} = 2\,\mathrm{Re}\left[r_{k1} A_{l1} \ln(z_1) + r_{k2} A_{l2} \ln(z_2)\right] \tag{8.6.5}$$

$$P_{lk} = 2\,\mathrm{Re}\left[q_{k1}(\mu_1 n_1 - n_2)\, A_{l1}/z_1 + q_{k2}(\mu_2 n_1 - n_2)\, A_{l2}/z_2\right] \tag{8.6.6}$$

In which the direction cosines are:

$$\begin{aligned} n_1 &= \cos(n, z_1) \\ n_2 &= \cos(n, z_2) \end{aligned} \tag{8.6.7}$$

Furthermore, the constants r and q are given by:

$$\begin{aligned} r_{11} &= a_{11}\mu_1^2 + a_{12} - a_{13}\mu_1 \\ r_{12} &= a_{11}\mu_2^2 + a_{12} - a_{13}\mu_2 \\ r_{21} &= a_{12}\mu_1 + a_{22}/\mu_1 - a_{23} \\ r_{22} &= a_{12}\mu_2 + a_{22}/\mu_2 - a_{23} \end{aligned} \tag{8.6.8}$$

and:

$$\begin{aligned} q_{11} &= \mu_1 \\ q_{12} &= \mu_2 \\ q_{21} &= -1 \\ q_{22} &= -1 \end{aligned} \tag{8.6.9}$$

Moreover, A_{lk} are series of complex constants, which are deduced from equilibrium and uniqueness of displacements around the singularity.

If we investigate the tractions on the boundary of a circular region around the singularity and equilibrate them with the internal load, then we obtain:

$$\mathrm{Im}\left[A_{l1}+A_{l2}\right]=-\frac{\Delta_{l2}}{4\pi}$$
$$\mathrm{Im}\left[\mu_1 A_{l1}+\mu_2 A_{l2}\right]=\frac{\Delta_{l1}}{4\pi},\ l=1,2 \qquad (8.6.10)$$

Therefore, by integrating the strain along a circular path around the singularity we will get the displacements.

Finally, one has:

$$\mathrm{Im}\left[r_{11}A_{l1}+r_{12}A_{l2}\right]=0$$
$$\mathrm{Im}\left[r_{21}A_{l1}+r_{22}A_{l2}\right]=0 \qquad (8.6.11)$$

Some Special Cases

The stress field near crack tips in an anisotropic solid can be divided into local modes of deformation, but the degree of simplifications achieved by this way is less than for the isotropic case.

Such a fact is valid because the crack surface displacements in general anisotropic bodies depend upon the directional properties of the material and do not necessarily occur in a planar fashion. Furthermore, in all the following discussions, the x and y axes are directed parallel and normal to the crack surface and the z-axis is taken to be parallel to the leading edge of a crack during deformation (see Figure 8.7.1).

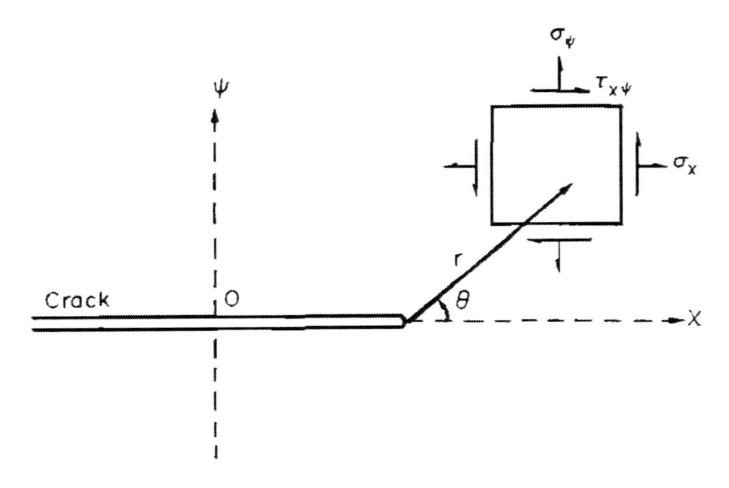

Figure 8.6.1. The x and y axis are directed parallel and normal to the crack surface and the z-axis is taken to be parallel to the leading edge of a crack during deformation.

For the symmetric loading the stress field near the crack tip has the following form: [17]

$$\begin{aligned}
\sigma_x &= \frac{K_1}{\sqrt{2r}}\operatorname{Re}\left[\frac{G_1G_2}{G_1-G_2}\left(\frac{G_2}{\sqrt{\cos\theta+G_2\sin\theta}}-\frac{G_1}{\sqrt{\cos\theta+G_1\sin\theta}}\right)\right]\\
\sigma_y &= \frac{K_1}{\sqrt{2r}}\operatorname{Re}\left[\frac{1}{G_1-G_2}\left(\frac{G_1}{\sqrt{\cos\theta+G_2\sin\theta}}-\frac{G_2}{\sqrt{\cos\theta+G_1\sin\theta}}\right)\right]\\
\tau_{xy} &= \frac{K_1}{\sqrt{2r}}\operatorname{Re}\left[\frac{G_1G_2}{G_1-G_2}\left(\frac{1}{\sqrt{\cos\theta+G_1\sin\theta}}-\frac{1}{\sqrt{\cos\theta+G_2\sin\theta}}\right)\right]
\end{aligned} \tag{8.6.12}$$

Also, for the skew-symmetric loading the stress field near the crack tip has the following formula: [17]

$$\begin{aligned}
\sigma_x &= \frac{K_2}{\sqrt{2r}}\operatorname{Re}\left[\frac{1}{G_1-G_2}\left(\frac{G_2^2}{\sqrt{\cos\theta+G_2\sin\theta}}-\frac{G_1^2}{\sqrt{\cos\theta+G_1\sin\theta}}\right)\right]\\
\sigma_y &= \frac{K_2}{\sqrt{2r}}\operatorname{Re}\left[\frac{1}{G_1-G_2}\left(\frac{1}{\sqrt{\cos\theta+G_2\sin\theta}}-\frac{1}{\sqrt{\cos\theta+G_1\sin\theta}}\right)\right]\\
\tau_{xy} &= \frac{K_2}{\sqrt{2r}}\operatorname{Re}\left[\frac{1}{G_1-G_2}\left(\frac{G_1}{\sqrt{\cos\theta+G_1\sin\theta}}-\frac{G_2}{\sqrt{\cos\theta+G_2\sin\theta}}\right)\right]
\end{aligned} \tag{8.6.13}$$

in whicch K_j, $j = 1, 2$ denote the stress-intensity factors.

As we can see from eqs. (8.6.12) and (8.6.13) the stress singularity at the crack tip is of order of $r^{-1/2}$ as in the isotropic case. On the contrary the angular distribution of the stresses, namely, the θ-dependence, depends upon the material properties. But the dependency of the crack tip stress distribution on remote boundary conditions is identical to the isotropic case, where the crack geometry and the applied loads affect only the intensity of stresses K_1.

REFERENCES

[1] S.G. Lekhnitskii, Theory of Elasticity of an Anisotropic Elastic Body, Holden-Day, San Francisco (1963).

[2] S.G. Lekhnitskii, Anisotropic Plates, Gordon and Breach, New York (1968).

[3] S.G. Lekhnitskii, Some cases of elastic equilibrium of the uniform cylinder with arbitrary anisotropy, *Prikl. Mat. Mekh.* 11, 61-359 (1939).

[4] S.G. Lekhnitskii, Stress distribution in rotating elliptical anisotropic plate, Uchen. Zap. Leningrgos. *Univ. ser Fiz.* - matem., No. 13 (1944).

[5] S.G. Lekhnitskii, Stress distribution in an anisotropic plate with an elliptic elastic core (plane problem), *Inz. Sb.* (1954).

[6] G.N. Savin, Stress Concentration Around Holes, Pergamon Press, New York (1965).

[7] G.N. Savin, Stresses in an anisotropic block at a given load on its surface (plane problem), *Vest. Inzh. Tekh.,* No.3 (1940).

[8] G.N. Savin, Pressure of an absolutely rigid punch of an elastic anisotropic medium (plane problem) *DAN UkrSSR*, No.2 (1939).

[9] G.N. Savin, Some problems of the theory of elasticity of an anisotropic medium, *Dokl. Akad. Nauk SSR* 23, (1939).

[10] G.N. Savin, Stress Concentration at Openings, Gostekhizdat, Moscow (1951).

[11] M.O. Basheleishvili, On the fundamental solutions of differential equations of an anisotropic elastic body, *Ibid.* 19, 393-400 (1957).

[12] M.O. Basheleishvili, Solution of plane boundary value problems of statics for an anisotropic elastic body, *Ibid.* 3, 93-139 (1963).

[13] J.R. Willis, Anisotropic elastic inclusion problems, *Q. J. Mech. Appl. Math.* 27, 157-174 (1964).

[14] H.T. Rathod, A study of asymmetrically loaded Griffith crack in an infinite anisotropic medium, *J. Engng Fract. Mech.* 11, 87-97 (1979).

[15] S. Krenk, The stress distribution in an infinite anisotropic plate with colinear cracks, *Int. J. Solids Struct.* 11, 449-460 (1975).

[16] G.C. Sih and H. Liebowitz, Mathematical theories of brittle fracture, in Fracture (Ed. H.Liebowitz), Vol.2, Academic Press, New York (1968).

[17] G.C. Sih and M.K. Kassir, Three-dimensional Crack Problems: Mechanics of Fracture, Vol.2, Noordhoff, Leyden, The Netherlands (1975).

[18] G.C. Sih, P.C. Paris and G.R. Irwin, On cracks in rectilinearly anisotropic bodies, *Int. J. Fracture Mech.* 1, 189-203 (1965).

[19] I.N. Sneddon, Fourier Transforms, McGraw-Hill, New York (1951).

[20] I.N. Sneddon, Integral transform methods for the solution of mixed boundary value problems in the classical theory of elastostatics, in Application of Integral Transforms in the Theory of Elasticity (Ed. I.N.Sneddon), CISM Courses and Lectures No.220, Springer-Verlag, New York (1975).

[21] G.E. Tupholme, A study of cracks in orthotropic crystals using dislocation layers, *J.Engng Math.* 8, 57-69 (1974).

[22] D.D. Ang and M.L. Williams, Combined stresses in an orthotropic plate having a finite crack, *J. Appl. Mech.* 28, 372-378 (1961).

[23] O.L. Bowie and C.E.Freese, Central crack in plane orthotropic rectangular sheet, *Int. J. Fracture Mech.* 8, 49-58 (1972).

[24] F.J. Rizzo and D.J.Shippy, A method for stress determination in plane anisotropic elastic bodies, *J. Compos. Mater.* 4, 36-61 (1970).

[25] S.M. Vogel and F.J. Rizzo, An integral equation formulation of three-dimensional anisotropic elastostatic boundary value problems, *J. Elasticity* 3, 203-216 (1973).

[26] M.D. Snyder and T.A. Cruse, Boundary-integral equation analysis of cracked anisotropic plates, *Int. J. Fracture* 11, 315-328 (1975).

[27] M.D. Snyder and T.A. Cruse, Crack tip stress intensity factors in finite anisotropic plates, Air Force Materials Laboratory, *Technical report AFML-TR*-73-209 (1973).

[28] E.G. Ladopoulos, On the solution of the two-dimensional problem of a plane crack of arbitrary shape in an anisotropic material, *J. Engng Fracture Mech.* 28, 187-195 (1987).

[29] E.G. Ladopoulos, Singular integral operators method for anisotropic elastic stress analysis, *Comp. Struct.* 48, 965-973 (1993).

[30] E.G. Ladopoulos, Singular Integral Equations, Linear and Non-linear Theory and its Applications in Science and Engineering, Springer-Verlag, Berlin, New York (2000).

[31] K.S. Parihar and S. Sowdamini, Stress distribution in a two-dimensional infinite anisotropic medium with collinear cracks, *J.Elasticity* 15, 193-214 (1985).

[32] T. Mura, Micromechanics of Defects in Solids, Martinus Nijhoff, The Hague (1982).

[33] C. Ouyang and Mei-Zi Lu, On a micromechanical fracture model for cracked reinforced composites, *Int. J. Non-Linear Mech.* 18, 71-77 (1983).

[34] R.P. Gilbert, G.C. Hsiao and M. Schneider, The two-dimensional linear orthotropic plate, Appl. Anal. 15, 147-169 (1983).

[35] R.P. Gilbert and M. Schneider, The linear anisotropic plate, *J. Compos. Mater.* 15, 71-78 (1981).

[36] R.P. Gilbert and R. Magnanini, The boundary integral method for two-dimensional orthotropic materials, *J. Elasticity* 18, 61-82 (1987).

[37] U. Zastrow, Solution of the anisotropic elastostatical boundary value problems by singular integral equations, *Acta Mech.* 44, 59-71 (1982).

[38] U. Zastrow, On the formulation of the fundamental solution for orthotropic plane elasticity, *Acta Mech.* 57, 113-128 (1985).

[39] U. Zastrow, On the complete system of fundamental solutions for anisotropic slices and slabs: a comparison by use of the slab analogy, *J. Elasticity* 15, 293-318 (1985).

[40] L.J. Gray, D. Ghosh and T. Kaplan, Evaluation of the anisotropic Green's function in three dimensional elasticity, *Comput. Mech.* 17, 255-261 (1996).

[41] M.A. Sales and L.J. Gray, Evaluation of the anisotropic Green's function and its derivatives, *Comp. Struct.* 69, 247-254 (1998).

[42] F. John, Plane Waves and Spherical Means Applied to Partial Differential Equations, Interscience, New York (1955).

Chapter 9

PLASTICITY OF ISOTROPIC SOLIDS BY MULTIDIMENSIONAL SINGULAR INTEGRAL EQUATIONS

9.1. INTRODUCTION

A big variety of plasticity problems of applied character are reduced to the solution of a system of multidimensional singular integral equations. Hence, such systems of integral equations must be numerically evaluated by using several computational recipes, as closed form solutions are not possible to be determined. These numerical recipes discretize the domain of the problem under consideration into a number of elements or cells. Then the governing equations of the problem are approximated over the region by functions which fully or partially satisfy the boundary conditions.

During the last years, several studies have been published on the application of the method of formulation of plasticity problems, by using systems of singular integral equations. J. L. Swedlow and T. A. Cruse [1], S. Mukherjee [2], A. Mendelson [3], H. D. Bui [4], J. C. F. Telles and C. A. Brebbia [5] and H. Poon, S. Mukherjee and M. F. Ahmad [6], studied and investigated some plasticity problems by using the Boundary Integral Equation Method (B.I.E.M.), or Boundary Element Method (B.E.M.), while E.G. Ladopoulos [7] - [12] has introduced and investigated the Singular Integral Operators Method (S.I.O.M.) for solving some basic plasticity problems.

Beyond the above, several other scientists have studied plasticity problems following theoretical or numerical classical lines, like finite-elements, etc. Among them we shall mention the following authors : O. C. Zienkiewicz et al. [13], G. C. Nayak and O. C. Zienkiewicz [14], W. F. Chen [15], Y. Yamada et al. [16], P. V. Marcal and I. P. King [17], R. Hill [18], O. Hoffman and G. Sachs [19], P. G. Hodge and G. N. White [20], W. T. Koiter [21], D. C. Drucker and W. Prager [22], W. Prager [23], D. N. Allen and R. V. Southwell [24], W. Johnson and P. B. Mellor [25], W. D. Liam Finn [26], Z. Mroz [27], J. N. Goodier and P. G. Hodge [28], J. H. Argyris [29], G. G. Pope [30], R. von Mises [31], J. Casey and P. M. Naghdi [32] - [34], P. M. Naghdi and J. A. Trapp [35], P. V. Lade [36], P. J. Yoder and W. D. Iwan [37], J. H. Prevost [38], and P. K. Banerjee and A. S. Stipho [39].

By the present chapter particular attention is concentrated on the construction and verification of algorithms which are effective in plasticity problems. Furthermore, the accuracy of the approximation, the number of iterations and other parameters of the algorithms must be closely related to the accuracy of the original data.

Moreover, it is sometimes necessary to solve a problem to a high degree of accuracy in successively different parts of the modelled domain. Under these conditions, finite element methods can become very uneconomical in terms of the required amounts of computer time or main storage owing to the generality of the gridding necessary to deal with the various loading conditions. On the other hand, the S.I.O.M. seems to be more economical when a high accuracy numerical technique is required for solving applied mechanics problems.

Further difficulties crop up in the course of modelling concentrated loads, since the solution then becomes infinite at the location of the load. The Singular Integral Operators Method surpasses these difficulties referring to plasticity problems and can be used for all kinds of modelling of concentrated loads with excellent accuracy.

The methods of numerical solution of the two- and three-dimensional singular integrals presented by the present chapter make it clear that the numerical solution of such equations is by no means impossible and in most cases it requires about the same amount of computational effort as the numerical solution of a regular Fredholm integral equation. Hence, several plasticity problems can be solved by reduction to a system of two- or three-dimensional singular integral equations, which combine accuracy and effectiveness. However, as the complexity of the plasticity problems to be solved increases, more tests must be carried out before the conditions under which the proposed method is advantageous, can be precisely defined.

In the present chapter are solved the following plasticity problems: Two-dimensional plastic stress analysis of isotropic solids, application to the determination of the plastic behaviour of a perforated tension strip in plane stress, application to the determination of the plastic behaviour of a thick sphere and a thick cylinder in plane strain, application of two-dimensional plasticity in a circular rocky tunnel, application to the determination of the plastic behaviour of a square block compressed by two opposite perfectly rough rigid punches in plane strain, three-dimensional plastic stress analysis of isotropic solids, fracture behaviour of double notched tensile specimen, three-dimensional thermoelastoplastic stress analysis of isotropic solids and thermoelastic stress analysis application for a hollow cylinder.

9.2. Linear Theory of Two-dimensional Plasticity of Isotropic Solids

The multidimensional singular integral equations combined with the numerical Singular Integral Operators Method (S.I.O.M.) are further used for the solution of basic problems of two-dimensional plastic stress analysis in isotropic solids. [8], [9]

Hence, consider the expression between the total strain rate $\dot{\varepsilon}_{ij}$ and the displacement rate $\dot{u}_i$:

$$\dot{\varepsilon}_{ij} = \frac{1}{2}\left(\dot{u}_{i,j} + \dot{u}_{j,i}\right) \tag{9.2.1}$$

Furthermore, the total strain rate is represented by the formula:

$$\dot{\varepsilon}_{ij} = \dot{\varepsilon}_{ij}^{e} + \dot{\varepsilon}_{ij}^{p} \tag{9.2.2}$$

where $\dot{\varepsilon}_{ij}^{e}$ and $\dot{\varepsilon}_{ij}^{p}$ are the elastic and plastic components of the strain rate tensor.

By applying Hooke's law to the elastic part of the strain rates, then the stress rates $\dot{\sigma}_{ij}$ reduce to the following form [1], [5], [9]:

$$\dot{\sigma}_{ij} = 2G\left(\dot{u}_{i,j} + \dot{u}_{j,i}\right) + \frac{2Gv}{1-2v}\dot{u}_{k,k}\delta_{ij} - 2G\dot{\varepsilon}_{ij}^{p} \tag{9.2.3}$$

in which v denotes Poisson's ratio and G the shear modulus.

From (9.2.1), (9.2.2) and (9.2.3) the stress rate tensor is assumed to be represented by:

$$\dot{\sigma}_{ij} = 2G\left(\dot{\varepsilon}_{ij} - \dot{\varepsilon}_{ij}^{p}\right) + \frac{2Gv}{1-2v}\left(\dot{\varepsilon}_{ll} - \dot{\varepsilon}_{kk}^{p}\right)\delta_{ij} \tag{9.2.4}$$

On the other hand, by inserting the plastic stress components $\dot{\sigma}_{ij}^{p}$ in eq. (9.2.4), one has:

$$\dot{\sigma}_{ij} = 2G\dot{\varepsilon}_{i,j} + \frac{2Gv}{1-2v}\dot{\varepsilon}_{ll}\delta_{ij} - \dot{\sigma}_{ij}^{p} \tag{9.2.5}$$

where:

$$\dot{\sigma}_{ij}^{p} = 2G\dot{\varepsilon}_{ij}^{p} + \frac{2Gv}{1-2v}\dot{\varepsilon}_{kk}^{p}\delta_{ij} \tag{9.2.6}$$

as is easily seen from (9.2.4) and (9.2.5).

Moreover, we consider the equilibrium conditions in the interior of the body:

$$\dot{\sigma}_{ij,i} + \dot{g}_{j} = 0 \tag{9.2.7}$$

in which $\dot{g}_{j}$ are the body force rates per unit volume.

On the boundary of the body, the equilibrium conditions are:

$$\dot{p}_{i} - \dot{\sigma}_{ij}\dot{n}_{j} = 0 \tag{9.2.8}$$

where $\dot{p}_{i}$ is the traction rate tensor per unit volume and n the outward normal to the boundary of the body.

Navier's equation for the two-dimensional problem is given by the following formula, after combining eqs. (9.2.1), (9.2.2), (9.2.4), (9.2.7) and (9.2.8):

$$\dot{u}_{j,ll} + \frac{1}{1-2v}\dot{u}_{l,lj} = -\frac{\dot{g}_{j}}{G} + 2\left(\dot{\varepsilon}_{ij,i}^{p} + \frac{v}{1-2v}\dot{\varepsilon}_{kk,j}^{p}\right), \quad (i, j, k, l = 1, 2) \tag{9.2.9}$$

A solution of Navier's equation (9.2.9) for the two-dimensional problem has the following form:

$$\dot{u}_i(y) = \int_{\Gamma} \left[U_{ij}(y, X)\dot{\tau}_j(X) - T_{ij}(y, X)\dot{u}_j(X) \right] \mathrm{d}\Gamma_x$$

$$+ \int_B U_{ij}(y,x)\dot{g}_j(x)\,\mathrm{d}B_x + \int_B 2GU_{ij,k}\dot{\varepsilon}^p_{jk}(x)\,\mathrm{d}B_x \tag{9.2.10}$$

$$(i, j, k = 1, 2)$$

where B is the area of cross-section of the body, Γ the boundary of B, τ_i the traction vector, X and Y are surface points and x and y are interior points.

Beyond the above, we consider the formula of symmetry of the plastic strain rate tensor:

$$2GU_{ij,k}\dot{\varepsilon}^p_{jk} = 2G\left[\frac{1}{2}\left(U_{ij,k} + U_{ik,j}\right)\right]\dot{\varepsilon}^p_{jk} \tag{9.2.11}$$

Recalling Kelvin's solution, eq. (9.2.11) takes the form:

$$\Sigma_{jki} = 2GU_{ij,k} = -\frac{1}{4\pi(1-v)r}\left[(1-2v)(\delta_{ij}r_{,k} + \delta_{ki}r_{,j}) - \delta_{jk}r_{,i} + 2r_{,i}r_{,j}r_{,k}\right] \tag{9.2.12}$$

where:

$$r = |x - y| \tag{9.2.13}$$

By using (9.2.12), then (9.2.10) becomes:

$$\dot{u}_i(y) = \int_{\Gamma} \left[U_{ij}(y, X)\dot{\tau}_j(X) - T_{ij}(y, X)\dot{u}_j(X) \right] \mathrm{d}\Gamma_x$$

$$+ \int_B U_{ij}(y,x)\dot{g}_j(x)\,\mathrm{d}B_x + \int_B \Sigma_{jki}(y,x)\dot{\varepsilon}^p_{jk}(x)\,\mathrm{d}B_x \tag{9.2.14}$$

$$(i, j, k = 1, 2)$$

The singular fundamental solutions corresponding to unit loads can be written as: [8]

$$U_{ij} = -\frac{1}{8\pi(1-v)G}\left[(3-4v)\ln(r)\delta_{ij} - r_{,i}r_{,j}\right] \tag{9.2.15}$$

$$T_{ij} = -\frac{1}{4\pi(1-v)r}\left[\left[(1-2v)\delta_{ij} + 2r_{,i}r_{,j}\right]\frac{\vartheta r}{\vartheta n} + (1-2v)(r_{,j}n_i - r_{,i}n_j)\right] \tag{9.2.16}$$

in which **n** denotes the outward normal to the boundary of the body.

By differentiating eq. (9.2.14), we obtain the stress rate tensor:

$$\dot{\sigma}_{ij}(y) = \int_{\Gamma} \left[-\Sigma^{1}_{ijk}(y,X)\dot{\tau}_k(X) - T_{ijk}(y,X)\dot{u}_k(X) \right] \mathrm{d}\Gamma_x$$

$$+ \int_{B} \left(-\Sigma^{1}_{ijk}(y,x) \right) \dot{g}_k(x)\, \mathrm{d}B_x - 2G\dot{\varepsilon}^{p}_{ij}(y) + \int_{B} \Sigma^{1}_{ijkl}(y,x)\dot{\varepsilon}^{p}_{kl}(x)\, \mathrm{d}B_x \tag{9.2.17}$$

where:

$$\Sigma^{1}_{ijk} = -\frac{1}{4\pi(1-v)r}\left[(1-2v)(\delta_{ij}r_{,k} + \delta_{ki}r_{,j} + \delta_{jk}r_{,i}) + 2r_{,i}r_{,j}r_{,k}\right] \tag{9.2.18}$$

which is Kelvin's solution:

$$\Sigma^{1}_{ijkl} = \Sigma_{ijkl} + \frac{G}{2\pi(1-2v)r^2}\left[4vr_{,i}r_{,j}\delta_{kl} - 2v\delta_{ij}\delta_{kl}\right] \tag{9.2.19}$$

where:

$$\Sigma_{ijkl} = \frac{G}{2\pi(1-v)r^2}\Big[2(1-2v)(\delta_{ij}r_{,k}r_{,l} + \delta_{kl}r_{,i}r_{,j}) + 2v(\delta_{li}r_{,j}r_{,k} + \delta_{jk}r_{,l}r_{,i} +$$

$$+ \delta_{ik}r_{,l}r_{,j} + \delta_{jl}r_{,i}r_{,k}) - 8r_{,i}r_{,j}r_{,k}r_{,l} + (1-2v)(\delta_{ik}\delta_{lj} + \delta_{jk}\delta_{li}) - (1-4v)\delta_{ij}\delta_{kl}\Big] \tag{9.2.20}$$

9.3. Two–Dimensional Application to the Determination of the Plastic Behaviour of a Perforated Tension Strip in Plane Stress

As an application of the previous theory we will determine the plastic behaviour of a perforated tension strip with strain hardening in plane stress (Figure 9.3.1). The same problem has been studied by O.C. Zienkiewicz et al. [13] by using one kind of finite element method, the Initial Stress Method, and by Y. Yamada et al. [16] and P.V. Marcal and I.P. King [17] by using another kind of finite element method, the Partial Stiffness Method. Moreover, this plane stress problem was also analysed experimentally by P.S. Theocaris and E. Marketos [40].

By the present study a comparison is presented between the new Singular Integral Operators Method (S.I.O.M.) [9], the two kinds of finite element (F.E.) methods, the Initial Stress Method (I.S.M.) and the Partial Stiffness Method (P.S.M.), and the Experimental Method (E.M.).

On the other hand, the perforated tension strip used has the following parameters: Young's modulus E=7000 kg/mm^2, Poisson's ratio v = 0.2 and uniaxial stress in the y

direction $\overline{\sigma}_y = 24.3kg/mm^2$. The geometrical sizes of the above strip are a = 20 *mm*, b = 18 *mm* and the diameter of the hole D = 10 *mm* (see: Figure 9.3.1).

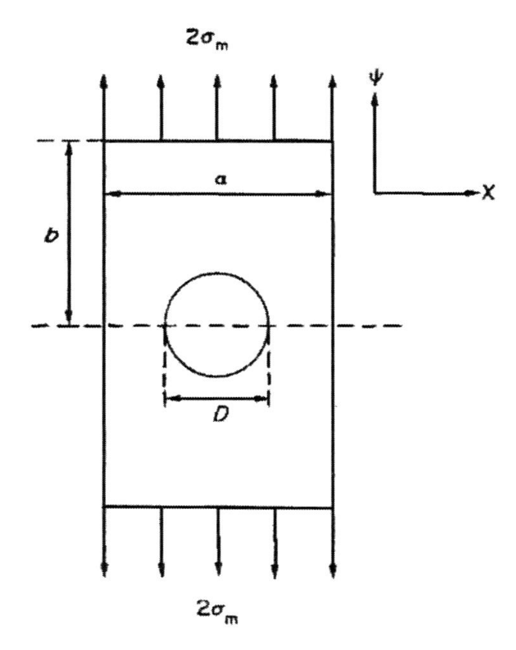

Figure 9.3.1. A perforated tension strip in plane stress.

Beyond the above, the slope of the curve relating the uniaxial stress H' is given by the formula:

$$H' = \mathrm{d}\overline{\sigma}/\mathrm{d}\,\varepsilon_{up} \tag{9.3.1}$$

where $\overline{\sigma}$ denotes the uniaxial yield stress and ε_{up} the plastic uniaxial strain, by using the von Mises yield criterion. On the other hand, in the case of the perforated tension strip of our application $H'/E = 0.032$ is valid.

For isotropic materials the uniaxial yield stress $\overline{\sigma}$ and the mean stress at the strip are given by the formulae:

$$\sigma_m = \sigma_{ii}/3 \tag{9.3.2}$$

$$\overline{\sigma} = \left[1/2 S_{ij} S_{ij}\right]^{1/2} \tag{9.3.3}$$

in which:

$$S_{ij} = \sigma_{ij} - 1/3\,\sigma_{ii}\delta_{ij} \tag{9.3.4}$$

where the stress tensor σ_{ij} is given by formula (9.2.17) and δ_{ij} denotes the function of Kronecker.

Hence, Figure 9.3.2 shows the relation between the load and the strain for the perforated tension strip in plane stress with strain hardening.

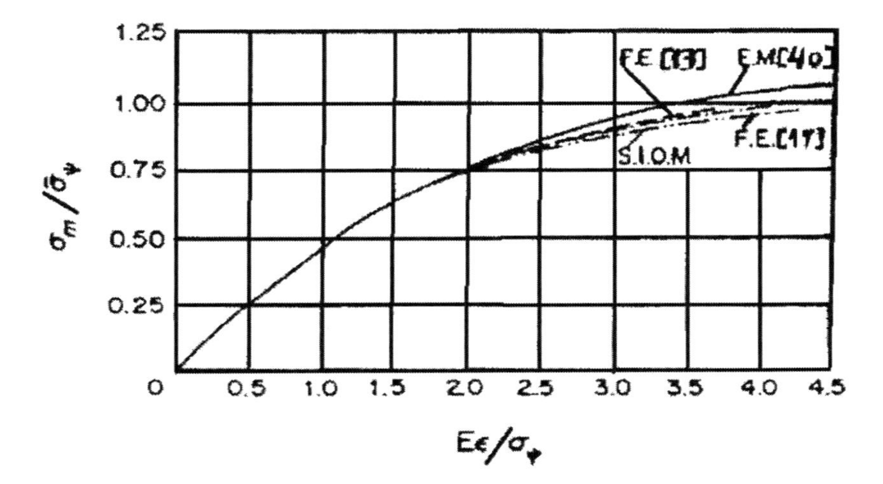

Figure 9.3.2. Relation between the load and the strain for a perforated tension strip in plane stress with strain hardening.

As it can be seen from Figure 9.3.2, the results of the Singular Integral Operators Method (S.I.O.M.) and the Initial Stress Method (I.S.M.) are closer to experimental results for the same load increments.

Figure 9.3.3 shows the relation between the uniaxial yield stress $\bar{\sigma}$ and the strain $\bar{\varepsilon}$ for the perforated tension strip in plane stress with strain hardening.

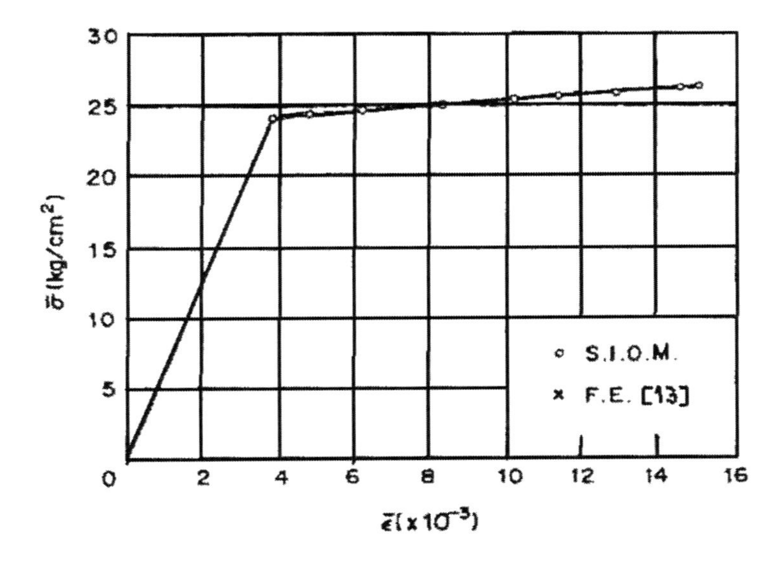

Figure 9.3.3. Relation between the uniaxial yield stress $\bar{\sigma}$ and the strain $\bar{\varepsilon}$ for a perforated tension strip in plane stress with strain hardening.

From Figure 9.3.3 it is seen that both the Singular Integral Operators and the Initial Stress Methods coincide very well.

The Initial Stress Method permits the advantage of initial processes to be retained. On the other hand, practice has shown that the Singular Integral Operators Method is often more economical than the two kinds of finite element methods, the Partial Stiffness Method and the Initial Stress Method.

9.4. TWO-DIMENSIONAL APPLICATION TO THE DETERMINATION OF THE PLASTIC STRESS ANALYSIS OF A THICK SPHERE AND A THICK CYLINDER IN PLANE STRAIN

As a second application of the linear theory of two-dimensional plasticity of isotropic solids we will determine the plastic behaviour of a thick sphere and a thick cylinder in plane strain (Figure 9.4.1). The same problem for the thick sphere was analysed by R. Hill [18] and by O. Hoffman and G. Sachs [19] by using finite elements.

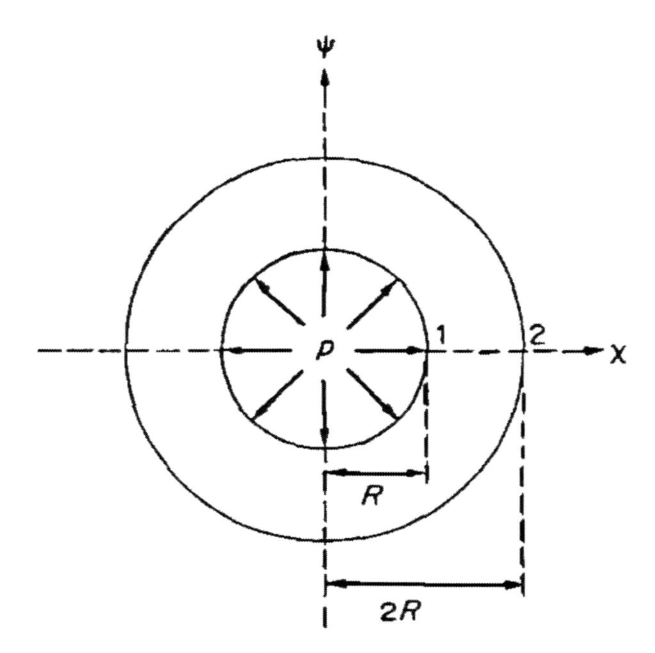

Figure 9.4.1. An ideal plastic thick sphere or an ideal plastic thick cylinder in plane strain.

Beyond the above, the problem of a thick cylinder in plane strain has been studied by R. Hill [18] using the Tresca finite element method and by P.G. Hodge and G.N. White [20] using the von Mises finite element method.

Thus, a comparison will be made between the new Singular Integral Operators Method (S.I.O.M.) [9] and the finite element methods.

The sphere and the cylinder used were ideal plastic ($H' = 0$) with Young's modulus $E = 10^7$ *psi*, Poisson's ratio $v = 0.33$ and uniaxial yield stress $\bar{\sigma}_0 = 10^4\, psi$. The radius of the sphere or the cylinder is $2R = 2\times 3$ *in*. (see Figure 9.4.1).

Consider the Tresca yield criterion: [14]

$$2/2^{1/2}\, S_{ij} \cos\theta - \overline{\sigma}_0 = 0 \tag{9.4.1}$$

where:

$$-\pi/6 \le \theta = 1/3 \sin^{-1}\left(-\frac{3\sqrt{3}J_3}{2\overline{\sigma}^3} \right) \le \pi/6 \tag{9.4.2}$$

$$J_3 = 1/3\, S_{ij} S_{jk} S_{ki} \tag{9.4.3}$$

and $\overline{\sigma}$ is given by eq. (9.3.3) and S_{ij} by eq. (9.3.4).

Moreover, the von Mises yield criterion can be written as following [14]:

$$(3/2)^{1/2} S_{ij} - \overline{\sigma}_0 = 0 \tag{9.4.4}$$

The Tresca and von Mises yield criteria are very well verified in metal plasticity.

Figure 9.4.2 shows the relation between the pressure and the displacement for the ideal plastic thick sphere. Furthermore, Figure 9.4.3 shows the relation between the pressure and the displacement for the ideal plastic thick cylinder. Finally, as it can be seen from Figs. 9.4.2 and 9.4.3, the results of the Singular Integral Operators Method and the finite element methods coincide very well.

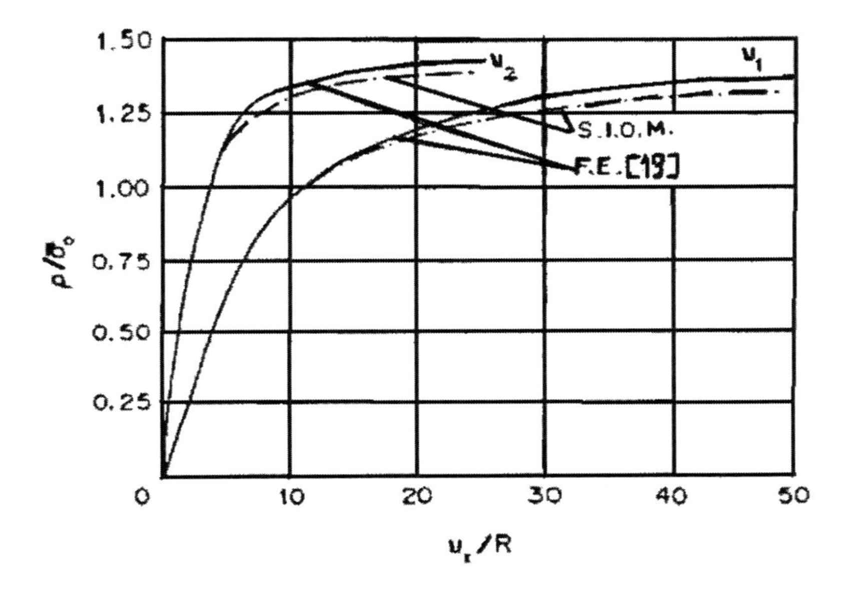

Figure 9.4.2. Relation between pressure and displacement for an ideal plastic thick sphere.

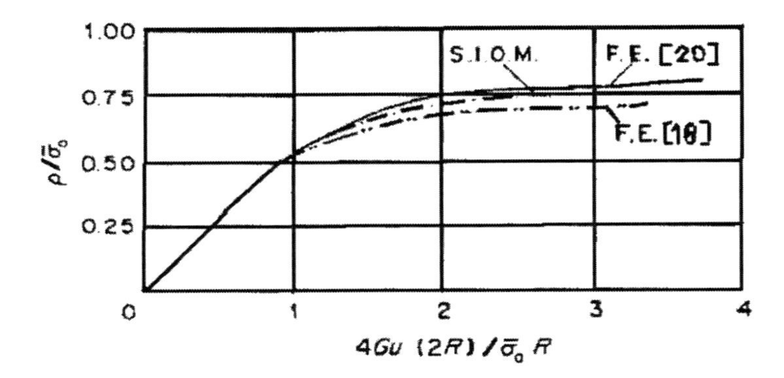

Figure 9.4.3. Relation between pressure and displacement for an ideal plastic thick cylinder.

9.5. Two-Dimensional Application of Plasticity in a Circular Rocky Tunnel

We will consider further a circular rocky tunnel studied by S.F. Reyes and D.U. Deere [41] with finite element methods. The results of this problem solved by the new proposed Singular Integral Operators Method (S.I.O.M.) are compared with the corresponding finite element results.

The rocky solid of the tunnel was assumed to be perfectly plastic with: Young modulus $E = 500$ *ksi*, Poisson's ratio $v = 0.2$, cohesion of the material $c = 0.28$ *ksi* and angle of internal friction $\varphi = 30^{\circ}$.

Beyond the above, the infinite domain of the material is initially subjected to a uniform stress field of 1 *ksi* vertical and 0.4 *ksi* horizontal. For the solution of this two-dimensional plasticity problem the Mohr-Coulomb yield criterion has been used under the Drucker-Prager simulation: [22]

$$c_1 \sigma_{ii} + (1/2)^{1/2} \left(\sigma_{ij} - 1/3 \, \sigma_{ii} \delta_{ij} \right) - c_2 = 0 \tag{9.5.1}$$

where:

$$c_1 = \tan \varphi / \left(9 + 12 \tan^2 \varphi \right)^{1/2} \tag{9.5.2}$$

$$c_2 = 3c / \left(9 + 12 \tan^2 \varphi \right)^{1/2} \tag{9.5.3}$$

in which σ_{ij} gives the stress tensor.

Therefore, by combining eqs. (9.5.1), (9.5.2) and (9.5.3) with eqn. (9.2.17), then the stress field was calculated. Figure 9.5.1 shows the stress tensor σ_x and σ_y by using the numerical technique described in the previous sections (S.I.O.M.) in comparison with the finite elements studied by S.F. Reyes and D.U. Deere [41].

The differences in σ_y values in Figure 9.5.1 are probably due to outer boundary conditions considered in the two different analyses, while for the σ_x values there are no differences between the computations of the two techniques.

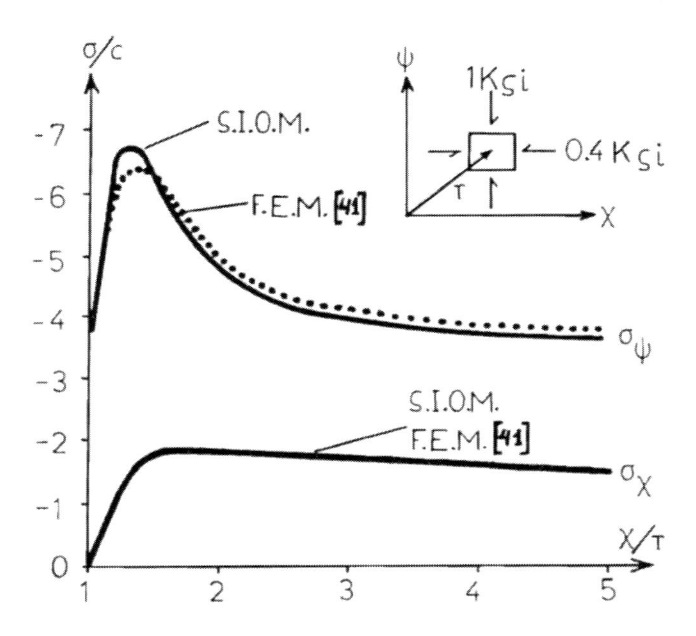

Figure 9.5.1. The stress tensor for a circular rocky tunnel by using S.I.O.M. and F.E.M.

9.6. Two-Dimensional Application to the Determination of the Plastic Stress Analysis of a Square Block Compressed by Two Opposite Perfectly Rough Rigid Punches in Plane Strain

In the present application we will determine the plastic behaviour of a square block compressed by two opposite perfectly rough rigid punches in plane strain, Figure 9.6.1. The same problem has been previously studied by W.F. Chen [15] and A.C.T. Chen and W.F. Chen [42] by using triangular Finite Elements. Hence, a comparison will be made between the Singular Integral Operators Method [10] and the Finite Element Method.

Furthermore, the material used was ideal plastic ($H' = 0$) by using the von Mises yield criterion, with Young modulus $E = 10^7$ *psi*, Poisson's ratio $v = 0.33$ and uniaxial yield stress $\overline{\sigma} = 13000\,psi$. The width of the punches is $d = 0.5$ *in* and the side of the square block $a = 3$ *in*, Figure 9.6.1.

For isotropic solids the uniaxial yield stress $\overline{\sigma}$ and the mean stress σ_m are given by the following relations:

$$\sigma_m = \sigma_{ii}/3 \tag{9.6.1}$$

$$\overline{\sigma} = \left[1/2\, S_{ij} S_{ij}\right]^{1/2} \tag{9.6.2}$$

with:

$$S_{ij} = \sigma_{ij} - 1/3\,\sigma_{ii}\delta_{ij} \tag{9.6.3}$$

in which the stress tensor σ_{ij} is given by eq. (9.2.17) and δ_{ij} denotes the function of Kronecker.

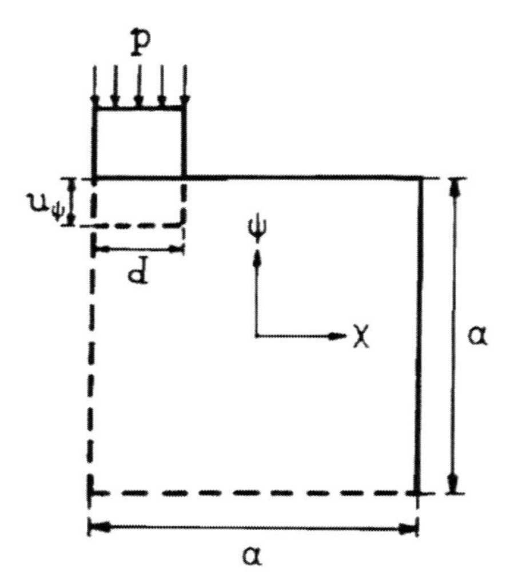

Figure 9.6.1. A square block compressed by two opposite perfectly rough rigid punches.

Figure 9.6.2 shows the relation between the mean pressure $\sqrt{3}\,p/2\overline{\sigma}$ and the applied displacement u_y/d, where the displacement u_y has been computed by eq. (9.2.14).

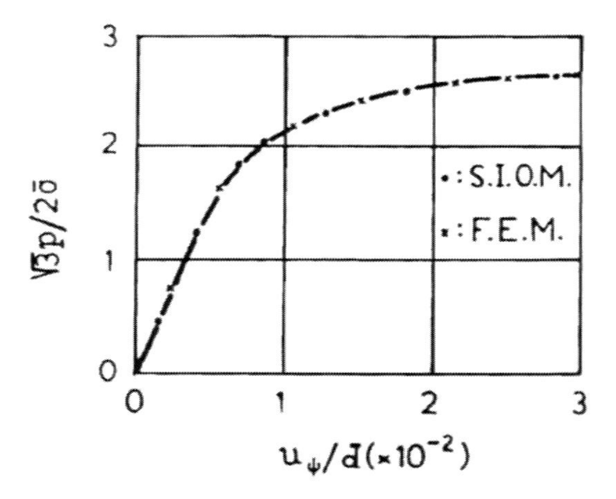

Figure 9.6.2. Relation between the mean pressure $\sqrt{3p}\ /2\,\overline{\sigma}$ and the applied displacement u_y/d.

From Figure 9.6.2 it can be seen that the results of the Singular Integral Operators Method and the Finite Element Method coincide very well. This example shows the advantages by using the S.I.O.M. in plasticity in preference to F.E.M.

The results of the present chapter demonstrate the application of the Singular Integral Operators Method to the two-dimensional elasto-plastic stress analysis for solving some important engineering problems of structural analysis. An application has been given, to the determination of the plastic behaviour of a square block compressed by two opposite perfectly rough rigid punches in plane strain.

In many practical applications it appears that the mentioned S.I.O.M. requires less execution time than the equivalent F.E.M. provided that the results are required at not too many interior points. The results on the surface are of primary interest, while results at interior points can be readily obtained by a judicious positioning of interfaces.

Beyond the above, a very interesting feature of the S.I.O.M. is the simplicity of the input data required to run a problem. This contrasts with the large amount of data needed to run a F.E. program. Such a property is very important in practice as many hours are lost in preparing and checking F.E. data.

Finally, the S.I.O.M. presents many advantages over "domain-type" techniques and these advantages are more evident for problems with boundaries at infinity.

9.7. Linear Theory of Three-Dimensional Plasticity of Isotropic Solids

The multidimensional singular integral equations in cooperation with the Singular Integral Operators Method (S.I.O.M.) are further used for the solution of three-dimensional plasticity problems for isotropic solids. [7]. The technique which will be presented makes use of the extended form of the Somigliana identity. Displacement rate can be expressed in terms of the traction and the displacement rate on the boundary including plastic strain rate. The resulting kernel has a strong singularity such that the gradient of the displacement rate may be evaluated by an integro-differential equation. The proposed method involves the use of initial stress, the initial strain and the fictitious forces. It may be regarded as an extension of the method of Fredholm integral equations and hence a direct application of the results of the potential theory. Numerical calculations can thus proceed in a straightforward fashion.

The total strain rate:

$$\dot{\varepsilon}_{ij} = \dot{\varepsilon}_{ij}^{e} + \dot{\varepsilon}_{ij}^{p} \tag{9.7.1}$$

consists of the sum of $\dot{\varepsilon}_{ij}^{e}$ and $\dot{\varepsilon}_{ij}^{p}$ that represent, respectively, the elastic and plastic components of the strain rate. The total strain rate and the displacement rate $\dot{u}_i$ are related as:

$$\dot{\varepsilon}_{ij} = \frac{1}{2}\left(\dot{u}_{i,j} + \dot{u}_{j,i}\right) \tag{9.7.2}$$

By application of the Hooke's law for the elastic part of the strain rate tensor, then the following expression for the stress rates is obtained:

$$\dot{\sigma}_{ij} = G\left(\dot{u}_{i,j} + \dot{u}_{j,i}\right) + \frac{2G\nu}{1-2\nu}\dot{u}_{k,k}\delta_{ij} - 2G\dot{\varepsilon}_{ij}^{p} \tag{9.7.3}$$

in which G denotes the shear modulus and ν the Poisson's ratio.

By making further use of eqs. (9.7.1) and (9.7.2), eq. (9.7.3) takes the form:

$$\dot{\sigma}_{ij} = 2G\left(\dot{\varepsilon}_{ij} - \dot{\varepsilon}_{ij}^{p}\right) + \frac{2Gv}{1-2v}\left(\dot{\varepsilon}_{ll} - \dot{\varepsilon}_{kk}^{p}\right)\delta_{ij} \tag{9.7.4}$$

In terms of the plastic stress components $\dot{\sigma}_{ij}^{p}$, the above result becomes:

$$\dot{\sigma}_{ij} = 2G\dot{\varepsilon}_{i,j} + \frac{2Gv}{1-2v}\dot{\varepsilon}_{ll}\delta_{ij} - \dot{\sigma}_{ij}^{p} \tag{9.7.5}$$

where the $\dot{\sigma}_{ij}^{p}$ is given by:

$$\dot{\sigma}_{ij}^{p} = 2G\dot{\varepsilon}_{ij}^{p} + \frac{2Gv}{1-2v}\dot{\varepsilon}_{kk}^{p}\delta_{ij} \tag{9.7.6}$$

The conditions of equilibrium are: [1], [5]

$$\dot{\sigma}_{ij,i} + \dot{a}_{j} = 0 \tag{9.7.7}$$

in which $\dot{a}_j$ stand for the body force rates per unit volume. On the other hand, on the boundary of the body, the equilibrium conditions are: [5]

$$\dot{p}_{i} - \dot{\sigma}_{ij}n_{j} = 0 \tag{9.7.8}$$

where n_j are components of the outward unit normal vector applied to the boundary of the body and $\dot{p}_i$ are the traction rates per unit area.

From eqs. (9.7.1), (9.7.2), (9.7.4), (9.7.7) and (9.7.8) the Navier's equations in three-dimensions are found:

$$\dot{u}_{j,ll} + \frac{1}{1-2v}\dot{u}_{l,lj} = -\frac{\dot{a}_j}{G} + 2\left(\dot{\varepsilon}_{ij,i}^{p} + \frac{v}{1-2v}\dot{\varepsilon}_{kk,j}^{p}\right) \tag{9.7.9}$$

a solution of which is: [4]

$$\dot{u}_i(y) = \int_S \left[U_{ij}(y,X)\dot{r}_j(X) - T_{ij}(y,X)\dot{u}_j(X)\right]\mathrm{d}S_x$$

$$+\int_V U_{ij}(y,x)\dot{a}_j(x)\,\mathrm{d}V_x + \int_V \Sigma_{jki}(y,x)\dot{\varepsilon}_{jk}^{p}(x)\,\mathrm{d}V_x \tag{9.7.10}$$

in which S is the surface, V the volume, τ_i the traction vector, X and Y are surface points and x and y are interior points.

By differentiating eq. (9.7.10) at a load point and by using eq. (9.7.3), the following formula for the stress rates is obtained:

$$\dot{\sigma}_{ij}(y) = \int_S \left[-\Sigma_{ijk}(y,X)\dot{\tau}_k(X) - T_{ijk}(y,X)\dot{u}_k(X) \right] \mathrm{d}S_x$$

$$+ \int_V \left(-\Sigma_{ijk}(y,x) \right) \cdot \dot{a}_k(x)\,\mathrm{d}V_x - \frac{G(8-10v)}{15(1-v)} \dot{\varepsilon}_{ij}^p(y) + \int_V \Sigma_{ijkl}(y,x)\dot{\varepsilon}_{kl}(x)\,\mathrm{d}V_x \qquad (9.7.11)$$

In eq. (9.7.11), Σ_{ijkl} are given by:

$$\Sigma_{ijkl} = \frac{G}{4\pi(1-v)r^3} \Big[3(1-2v)(\delta_{ij} r_{,k} r_{,l} + \delta_{kl} r_{,i} r_{,j}) +$$
$$+3v(\delta_{li} r_{,j} r_{,k} + \delta_{jk} r_{,l} r_{,i} + \delta_{ik} r_{,l} r_{,j} + \delta_{jl} r_{,i} r_{,k})$$
$$-15 r_{,i} r_{,j} r_{,k} r_{,l} + (1-2v)(\delta_{ik}\delta_{lj} + \delta_{jk}\delta_{li}) - (1-4v)\delta_{ij}\delta_{kl} \Big] \qquad (9.7.12)$$

where:

$$T_{ijk} = \Sigma_{ijk} n_l \qquad (9.7.13)$$

$$r = |x - y| \qquad (9.7.14)$$

Fundamental Solutions

In accordance with the method in [43] for the elliptic systems of linear partial differential equations, the delta function for scalars can be introduced:

$$\delta(r) = -\frac{1}{8\pi^2} \Delta_y \int_{|\zeta|=1} \delta(\mathbf{r} \cdot \zeta)\,\mathrm{d}\omega_\zeta \qquad (9.7.15)$$

with Δy being the Laplacean with respect to y_i. With the aid of eq. (9.7.15), a fundamental solution is equal to:

$$U_{ij}(x,y) = \frac{1}{8\pi^2} \Delta_y \int_{|\zeta|=1} V_{ij}(x,y,\zeta)\,\mathrm{d}\omega_\zeta \qquad (9.7.16)$$

where $V_{ij}(x,y,\zeta)$ is given by:

$$V_{ij}(x,y,\zeta) = \begin{cases} u_{ij}(x,y,\zeta) = A_{ij}(\zeta)(x_k - y_k)\zeta_k, & (x-y)\cdot\zeta > 0 \\ 0, & (x-y)\cdot\zeta \le 0 \end{cases} \qquad (9.7.17)$$

in which:

$$A_{ij}(\zeta) = \frac{\frac{1}{2}\varepsilon_{imn}\varepsilon_{jrs}B_{mr}(\zeta)Q_{ns}(\zeta)}{\det B} \tag{9.7.18}$$

and:

$$B_{ik}(\zeta) = C_{ijkl}\zeta_j\zeta_l \tag{9.7.19}$$

Thus, from eqs. (9.7.16) and (9.7.17), a fundamental solution of the form:

$$U_{ij}(x,y) = \frac{1}{8\pi^2}\Delta_y \int_{\substack{|\zeta|=1 \\ r\cdot\zeta=0}} A_{ij}(\zeta)(x_k - y_k)\zeta_k \, \mathrm{d}\omega_\zeta \tag{9.7.20}$$

is obtained. Here, the Laplacean Δ_y is given by:

$$\Delta_y \cdot r = \frac{2}{r} \tag{9.7.21}$$

Hence, eqs. (9.7.20) and (9.7.21) reduce to:

$$U_{ij}(x,y) = \frac{1}{8\pi^2 r}\oint_{|\zeta|=1} A_{ij}(\zeta)\,\mathrm{d}s \tag{9.7.22}$$

where ds is an element of arc length.

For the special case of an isotropic solid, c_{ijkl} is given by:

$$c_{ijkl} = \frac{2Gv}{(1-2v)}\delta_{ij}\delta_{kl} + G\left(\delta_{ik}\delta_{jl} + \delta_{il}\delta_{jk}\right) \tag{9.7.23}$$

Moreover, eqs. (9.7.18), (9.7.19) and (9.7.23) can be used to yield:

$$A_{ij} = \frac{1}{G}\left[\delta_{ij} - \frac{1}{2(1-v)}\zeta_i\zeta_j\right]\mathrm{d}s \tag{9.7.24}$$

Consider now the direction cosines of r to be denoted by β_{ij}. These result:

$$\zeta_i = \beta_{ij}\lambda_j \tag{9.7.25}$$

where λ_j is the new axes, so that the plane $\lambda_3 = 0$ is perpendicular to the vector **r**. With the aid of eq. (9.7.24), then eq. (9.7.22) becomes:

$$U_{ij}(x,y) = \frac{1}{8\pi^2 Gr} \oint\limits_{|\zeta|=1} \left[\delta_{ij} - \frac{1}{2(1-v)} \zeta_i \zeta_j \right] \mathrm{d}s \tag{9.7.26}$$

Further reduction as a result of eq. (9.7.25) renders:

$$U_{ij}(x,y) = \frac{1}{8\pi^2 Gr} \int_0^{2\pi} \left[\delta_{ij} - \frac{1}{2(1-v)} \beta_{ik} \beta_{jl} n_k n_l \right] \mathrm{d}\varphi \tag{9.7.27}$$

where:

$$n_1 = \cos\varphi, \quad n_2 = \sin\varphi \quad and \quad n_3 = 0$$

Therefore, equation (9.7.27) can be integrated to become:

$$U_{ij} = \frac{1}{8\pi Gr} \left[\left(2 - \frac{1}{2(1-v)} \right) \delta_{ij} + \frac{1}{2(1-v)} \beta_{i3} \beta_{j3} \right] \tag{9.7.28}$$

The three-dimensional fundamental solution finally can be written as:

$$U_{ij} = \frac{1}{16\pi(1-v)Gr} \left[(3-4v)\delta_{ij} + r_{,i} r_{,j} \right] \tag{9.7.29}$$

where:

$$\delta_{ij} = \beta_{ki} \beta_{kj} \tag{9.7.30}$$

The corresponding boundary tractions of eq. (9.7.27) are:

$$T_{im} = c_{ijkl} U_{km,l} n_j \tag{9.7.31}$$

By means of eq. (9.7.23), eq. (9.7.31) becomes:

$$T_{ij} = \frac{2Gv}{(1-2v)} U_{ki,k} n_i + G(U_{ij,k} + U_{kj,i}) n_k \tag{9.7.32}$$

Inserting eq. (9.7.29) into (9.7.32) follows the relation:

$$T_{ij} = -\frac{(1-2v)}{8\pi(1-v)r^2} \left[\left(\delta_{ij} + \frac{3}{(1-2v)} \beta_{i3} \beta_{j3} \right) \beta_{k3} n_k + \beta_{i3} n_j - \beta_{j3} n_i \right] \tag{9.7.33}$$

This is the expression for the boundary tractions in three dimensions.

9.8. THREE-DIMENSIONAL APPLICATION OF A DOUBLE NOTCHED TENSILE SPECIMEN

In order the three-dimensional plastic stress analysis to be applied, consider the fracture behaviour of an ideal plastic notched tensile specimen as shown in Figure 9.8.1. The same problem for the case of plane stress case has been analysed by G.C. Nayak and O.C. Zienkiewicz [14], Y. Yamada, N. Yoshimura and T. Sakurai [16] and D.N. Allen and R.V.Southwell [24]. by using the finite element method.

The corresponding case of plane strain was considered by W.F. Chen [15], P.V. Marcal and I.P. King [17] and O.C. Zienkiewicz, S. Valiappan and I.P. King [13]. In everything what follows, a comparison between the SIOM [7] and Finite Element (FE) will be made.

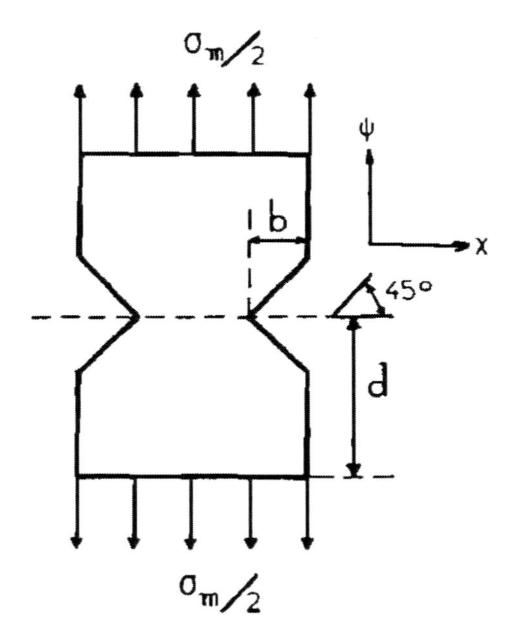

Figure 9.8.1. An ideal plastic notched tensile specimen.

The double notched tensile specimen is made of an ideal plastic material with Young modulus E = 7000 kg/mm^2, Poisson's ratio $\nu = 0.2$ and uniaxial stress $\overline{\sigma} = 24.3 kg/mm^2$. The geometrical sizes of the above specimen are b = 5 mm and d = 18 mm as specified in Figure 9.8.1.

For an isotropic solid, the uniaxial yield stress $\overline{\sigma}$ and the mean stress at the notch σ_m are equal to: [14]

$$\sigma_m = \frac{1}{3}\sigma_{ii} \tag{9.8.1}$$

and:

$$\bar{\sigma} = \sqrt{\frac{1}{2} S_{ij} S_{ij}} \tag{9.8.2}$$

where:

$$S_{ij} = \sigma_{ij} - \frac{1}{3}\sigma_{kk}\delta_{ij} \tag{9.8.3}$$

The stress tensor σ_{ij} is given by eq. (9.7.11) and δ_{ij} denotes the Kronecker delta function.

Figure 9.8.2 shows the relation between the load and the displacement of the specimen in Figure 9.8.1 under plane stress.

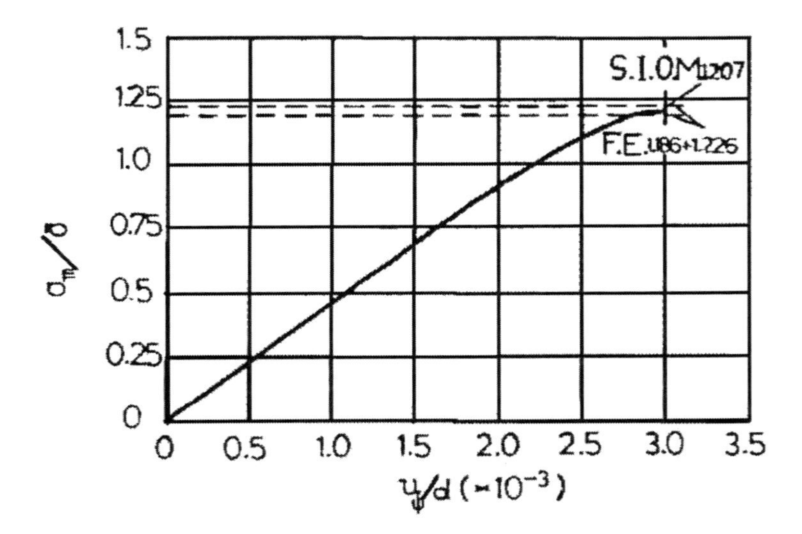

Figure 9.8.2. Relation between the load and the displacement for an ideal plastic notched tensile specimen in plane stress.

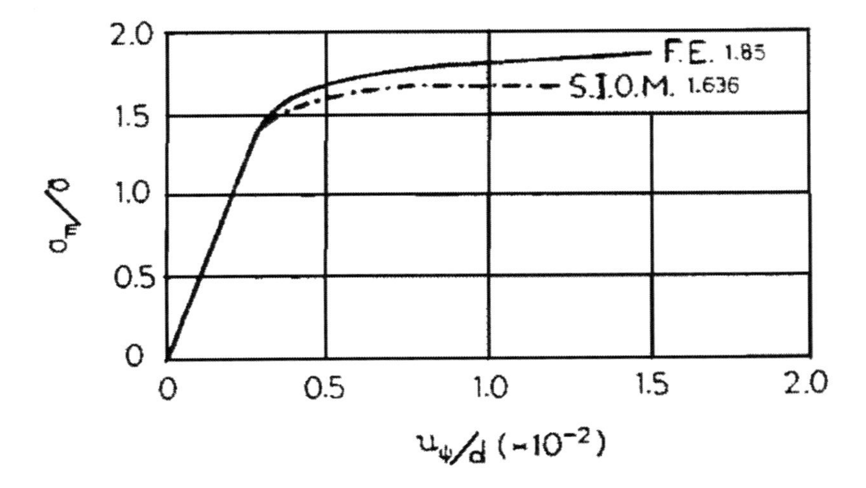

Figure 9.8.3. Relation between the load and the displacement for an ideal plastic notched tensile specimen in plane strain.

The normalized limit $\sigma_m/\overline{\sigma}$ estimated from the SIOM is about 1.207. On the other hand, the corresponding limit obtained by using the triangular and linear quadrilateral elements [14] is between 1.186 and 1.226. For the parabolic and cubic elements, the plastic zones developed more rapidly and collapse load occurred at 1.186.

Figure 9.8.3 shows the relation between the load and the displacement for the notched specimen. The ratio of $\sigma_m/\overline{\sigma}$ equal to 1.636 is obtained by the SIOM. A corresponding limit load of 1.850 has been found by Chen [15] using finite elements.

9.9. Linear Theory of Three-Dimensional Thermoelastoplastic Analysis of Isotropic Solids

Let us consider the total strain rate $\dot{\varepsilon}_{ij}$, which is the sum of the elastic $\dot{\varepsilon}_{ij}^{e}$, the plastic $\dot{\varepsilon}_{ij}^{p}$ and the thermal $\dot{\varepsilon}_{ij}^{T}$ strains: [11]

$$\dot{\varepsilon}_{ij} = \dot{\varepsilon}_{ij}^{e} + \dot{\varepsilon}_{ij}^{p} + \dot{\varepsilon}_{ij}^{T}, \quad i,j = 1,2,3 \tag{9.9.1}$$

in which:

$$\dot{\varepsilon}_{ij}^{T} = a\dot{T}\delta_{ij} \tag{9.9.2}$$

and where a is the coefficient of linear thermal expansion, T the temperature and δ_{ij} Kronecker's delta.

By assuming the plastic strains $\dot{\varepsilon}_{ij}^{p}$ to be deviatoric we have:

$$\dot{\varepsilon}_{ii}^{p} = \dot{\varepsilon}_{11}^{p} + \dot{\varepsilon}_{22}^{p} + \dot{\varepsilon}_{33}^{p} = 0 \tag{9.9.3}$$

Beyond the above, the relation between the total strains $\dot{\varepsilon}_{ij}$ and displacements $\dot{u}_i$ is equal to:

$$\dot{\varepsilon}_{ij} = \left(\dot{u}_{i,j} + \dot{u}_{j,i}\right)/2 \tag{9.9.4}$$

Thus, by combining eqs. (9.9.1) - (9.9.4), Hooke's law and the equations of equilibrium, we obtain Navier's equation for the three-dimensional thermoelastoplastic problem:

$$\dot{u}_{i,jj} + \frac{1}{1-2v}\dot{u}_{k,ki} = -\frac{\dot{B}_i}{G} + 2\dot{\varepsilon}_{ij,j}^{p} + \frac{2(1+v)}{1-2v}a\dot{T}_{,i} \tag{9.9.5}$$

where v denotes Poisson's ratio, G the shear modulus and B_i the prescribed body force per unit volume.

A solution of the Navier's equation (9.9.5) is given by the Somigliana identity: [2],[4]

$$\dot{u}_i(x) = \int_S \left[U_{ij}(x,Y)\dot{\tau}_j(Y) - T_{ij}(x,Y)\dot{u}_j(X)\right] \mathrm{d}\, S_Y$$
$$+ \int_V U_{ij}(x,y)\dot{B}_j(y)\,\mathrm{d}\,V_y + \int_V \Sigma_{jki}(x,y)\left[\dot{\varepsilon}^p_{jk}(y) + \delta_{jk} a\dot{T}(y)\right]\mathrm{d}\,V_y \tag{9.9.6}$$

in which V denotes the volume of the body, S its surface, X and Y are surface points, x and y interior points and $\dot{\tau}_i$ the traction vector:

$$\dot{\tau}_i = \dot{\sigma}_{ij} m_j \tag{9.9.7}$$

with $\dot{\sigma}_{ij}$ the stress rate tensor and m_j the outward unit normal to the surface S.

In eq. (9.9.6) $U_{ij}(x,y)$ denotes the Kelvin-Somigliana tensor:

$$U_{ij}(x,y) = \frac{1}{16\pi G(1-v)r}\left\{(3-4v)\delta_{ij} + \frac{(x_i - y_i)(x_j - y_j)}{r^2}\right\} \tag{9.9.8}$$

$\Sigma_{jki}(x,y)$ is the stress tensor corresponding to $U_{ij}(x,y)$:

$$\Sigma_{jki}(x,y) = -\frac{1}{8\pi(1-v)r^2}\begin{bmatrix} (1-2v)\left(\delta_{jk}\dfrac{\vartheta r}{\vartheta y_i} + \delta_{ji}\dfrac{\vartheta r}{\vartheta y_k} - \delta_{ki}\dfrac{\vartheta r}{\vartheta y_j}\right) + \\ +3\dfrac{\vartheta r}{\vartheta y_i}\dfrac{\vartheta r}{\vartheta y_k}\dfrac{\vartheta r}{\vartheta y_i} \end{bmatrix} \tag{9.9.9}$$

and T_{ij} the traction vector corresponding to Σ_{jki}:

$$T_{ij} = \Sigma_{jki} m_k \tag{9.9.10}$$

with r the distance between the points x, y:

$$r = |\mathbf{x} - \mathbf{y}| \tag{9.9.11}$$

Moreover, the stress rate tensor $\dot{\sigma}_{ij}$ is valid as:

$$\dot{\sigma}_{ij} = G(\dot{u}_{i,j} + \dot{u}_{j,i}) + \frac{2Gv}{1-2v}\dot{u}_{k,k}\delta_{ij} - 2\left(G\dot{\varepsilon}^p_{ij} + G\left(\frac{1+v}{1-2v}\right)a\dot{T}\delta_{ij}\right) \tag{9.9.12}$$
$$i, j, k = 1,2,3$$

Hence, by differentiating Somigliana's identity (9.9.6) at a load point and using (9.9.12) we obtain:

$$\dot{\sigma}_{ij}(x) = -\int_S \left[\Sigma_{ijk}(x,Y)\dot{\tau}_k(Y) + T_{ijk}(x,Y)\dot{u}_k(Y)\right] \mathrm{d}S_Y$$

$$-\int_V \Sigma_{ijk}(x,y)\dot{B}_k(y)\,\mathrm{d}V_y + 2G\dot{\varepsilon}_{ij}^p(x) + 3Ka\dot{T}(x)\delta_{ij}$$

$$+\int_V \Sigma_{ijkn}(x,y)\left[\dot{\varepsilon}_{kn}^p(y) + \delta_{kn}a\dot{T}(y)\right]\mathrm{d}V_y, \qquad i,j,k,n = 1,2,3 \tag{9.9.13}$$

where K denotes the bulk modulus and Σ_{ijkn} is valid as:

$$\Sigma_{ijkn} = \frac{G}{4\pi(1-v)r^3}\left[3(1-2v)(\delta_{ij}r_{,k}r_{,n} + \delta_{kn}r_{,i}r_{,j}) + \right.$$
$$3v(\delta_{ni}r_{,j}r_{,k} + \delta_{jk}r_{,n}r_{,i} + \delta_{ik}r_{,n}r_{,j} + \delta_{jn}r_{,i}r_{,k}) - 15r_{,i}r_{,j}r_{,k}r_{,n} + \tag{9.9.14}$$
$$\left.(1-2v)(\delta_{ik}\delta_{nj} + \delta_{jk}\delta_{ni}) - (1-4v)\delta_{ij}\delta_{kn}\right]$$
$$i,j,k,n = 1,2,3$$

In eq. (9.9.13) T_{ij} denotes the traction vector corresponding to Σ_{ijkn}:

$$T_{ijk} = \Sigma_{ijkn}m_n \tag{9.9.15}$$

where m_n denotes the outward unit normal to the surface S.

Finally, the stress tensor (9.9.13) gives the complete three-dimensional thermoelastoplastic stress analysis for every isotropic body.

9.10. Three-Dimensional Thermoelastoplastic Stress Analysis Application for a Hollow Cylinder

As an application of the previously mentioned thermoelastoplastic theory, we consider a hollow cylinder of height l with inner and outer radii of R_1 and R_2, respectively (see Figure 9.10.1).

The cylinder is subjected to the temperature distribution:

$$T(x_3) = \left(1/l^2\right)\left[l^2 - x_3^2 + lx_3\right], \quad 0 \le x_3 \le 1 \tag{9.10.1}$$

and its top and bottom surfaces are traction free. Beyond the above, the lateral surfaces of the cylinder are subject to a specific traction distribution with zero net force and moment.

The same problem has been previously theoretically solved by B.A. Boley and Y.H. Weiner [44, p. 278]. In this chapter the general three-dimensional thermoelastoplastic displacement distribution given by (9.9.6) is used in order to determine the radial and axial displacements. Moreover, the three-dimensional thermoelastoplastic stress tensor (9.9.13) is used to determine the transverse stresses.

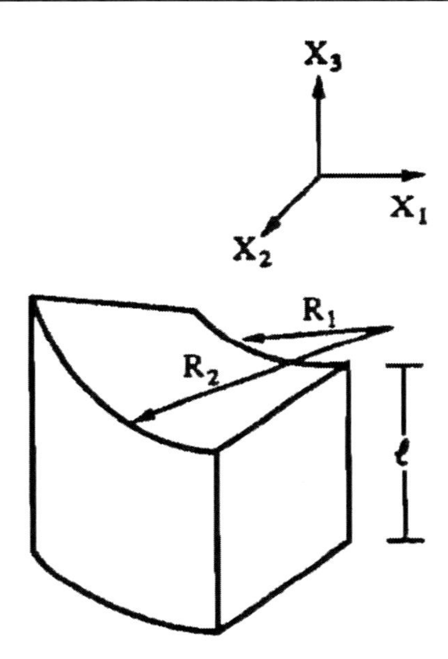

Figure 9.10.1. A hollow cylinder of height l with inner and outer radii of R_1 and R_2, respectively.

For the numerical evaluation of the two-dimensional singular integral (9.9.9) the Gauss-Legendre cubature formula is used [7], [8]. Consider therefore the singular integral:

$$\varphi(x_0, y_0) = \int_S w(x, y) \frac{f(x_0, y_0, \theta)}{r^2} u(x, y) \, \mathrm{d}S \tag{9.10.2}$$

where:

$$(x - x_0) + i(y - y_0) = re^{i\theta} \tag{9.10.2a}$$

and $w(x,y)$ is the weight function.

Using the Gauss-Legendre cubature formula, the weight function is equal to the unit and the singular integral (9.10.2) is approximated as follows: [7]

$$\varphi \cong \sum_{k=1}^{m} R_m \frac{\left[\varphi_A(x_0, y_0, x_m) - \varphi_B(x_0, y_0, x_m)\right]}{x_m - x_0} - 2\left[\varphi_A(x_0, y_0) - \varphi_B(x_0, y_0)\right] Z_n(x_0) \tag{9.10.3}$$

where the functions φ_A and φ_B are equal to:

$$\varphi_A \cong \sum_{k=1}^{n} R_k \frac{f(x_0, y_0, x, y_k)}{y_k - y_0} u(x, y_k) - 2f(x_0, y_0, x) u(x, y) Z_n(y_0), \tag{9.10.4}$$

$$y_0 \neq y_k, \qquad k = 1,2,\ldots,n$$

and:

$$\varphi_B \cong \sum_{\substack{k=1 \\ k\neq m}}^{n} R_k \frac{f(x_0, y_0, x, y_k)}{y_k - y_0} u(x, y_k) + R_m \left. \frac{\mathrm{d}[f(x_0, y_0, x, y)u(x, y)]}{\mathrm{d}\, y} \right|_{y=y_0}$$

$$- 2f(x_0, y_0, x)u(x, y_0)L_n(y_0),$$
$$y_0 = y_m, \quad k = 1,2,\ldots,n \tag{9.10.5}$$

with:

$$Z_n(x_0) = Q_n(x_0)/P_n(x_0) \tag{9.10.6}$$

$$L_n(y_0) = \frac{1}{P_n'(y_0)} \left[Q_n'(y_0) + \frac{1}{4} R_m P_n''(y_0) \right] \tag{9.10.7}$$

in which R_m are the weights and $P_n(x_0)$ and $Q_n(x_0)$ are the Legendre polynomials of degree n and the Legendre function of the second kind and order n, respectively.

On the other hand, for the numerical evaluation of the three-dimensional singular integral (9.9.14), the following method will be used [8], [11]. Consider, the three-dimensional singular integral defined on a three-dimensional finite region V, containing the third-order pole (x_0, y_0, z_0):

$$\varphi(x_0, y_0, z_0) = \int_V \frac{f(x_0, y_0, z_0, \theta, \varphi)}{r^3} u(x, y, z) \mathrm{d}V$$

$$= \int_V \frac{f(x_0, y_0, z_0, \theta, \varphi)}{\left[(x - x_0)^2 + (y - y_0)^2 + (z - z_0)^2\right]^{3/2}} u(x, y, z) \mathrm{d}V \tag{9.10.8}$$

Consider the following system of spherical coordinates:

$$\begin{aligned} x &= x_0 + r\sin\theta\cos\varphi, \quad \theta \in [0, \pi] \\ y &= y_0 + r\sin\theta\sin\varphi, \quad \varphi \in [0, 2\pi] \\ z &= z_0 + r\cos\theta \end{aligned} \tag{9.10.9}$$

Hence, the three-dimensional singular integral (9.10.8) reduces to:

$$\varphi(r, \theta, \varphi) = \lim_{\varepsilon \to 0} \int_0^{\pi} \int_0^{2\pi} \int_{\varepsilon}^{R(\theta,\varphi)} \sin\theta \, f(\theta, \varphi) \frac{u(r, \theta, \varphi)}{r} \mathrm{d}\theta \, \mathrm{d}\varphi \, \mathrm{d}r \tag{9.10.10}$$

By using the trapezoidal rule with *M, N* abscissae, we obtain the formula:

$$\varphi = \frac{2\pi^2}{MN}\sum_{i=1}^{M}\sum_{j=1}^{N}\sin\theta_i\varphi_j(\theta_i,\varphi_i) \qquad (9.10.11)$$

where:

$$\theta_i = \frac{\pi}{M}(i-1) \qquad (9.10.12a)$$

$$\varphi_j = \frac{2\pi(j-1)}{N} \qquad (9.10.12b)$$

$$\varphi(\theta,\varphi) = f(\theta,\varphi)\int_0^{R(\theta,\varphi)}\frac{u(r,\theta,\varphi)}{r}\mathrm{d}r \qquad (9.10.12c)$$

where the one-dimensional singular integral (9.10.12c) is numerically evaluated as:

$$\int_0^{R(\theta,\varphi)}\frac{u(r,\theta,\varphi)}{r}\mathrm{d}r \cong \sum_{k=1}^{B}D_k u\left[R(\theta,\varphi)\rho_k,\theta,\varphi\right] + u(0,\theta,\varphi)\ln\left[R(\theta,\varphi)\right] \qquad (9.10.13)$$

in which ρ_k are the abscissae and D_k the weights for the integration interval [0,1].

Consider further the height of the hollow cylinder of $l = 1.0\ m$ and its radii $R_1 = 0.4\ m$ and $R_2 = 1.0\ m$. Tables 9.10.1 - 9.10.3 show the relation between the radial, axial displacements and transverse stresses, respectively, and the parameters R, $R_1 \le R \le R_2$ and x_3, by using SIOM and the exact solution given in [44].

Table 9.10.1. Radial displacements distribution in the hollow cylinder of Figure 9.10.1

Coordinates		Radial displacements	
x_3	R	SIOM	Exact
0.00	0.40	0.2663	0.267
	0.70	0.4667	0.467
	1.00	0.6693	0.667
0.25	0.40	0.3622	0.367
	0.70	0.6415	0.642
	1.00	0.9162	0.917
0.50	0.40	0.4654	0.467
	0.70	0.8112	0.817
	1.00	0.1663	1.167
0.70	0.40	0.5664	0.567
	0.70	0.9918	0.992
	1.00	1.4166	1.417
1.00	0.40	0.6663	0.667
	0.70	1.1667	1.167
	1.00	1.6665	1.667

Table 9.10.2. Axial displacements distribution in the hollow cylinder of Figure 9.10.1

Coordinates		Axial displacements	
X_3	R	SIOM	Exact
0.00	0.40	-0.08031	-0.080
	0.70	-0.24512	-0.245
	1.00	-0.50042	-0.500
0.25	0.40	0.24796	0.248
	0.70	0.08314	0.083
	1.00	-0.17221	-0.172
0.50	0.40	0.58682	0.587
	0.70	0.42204	0.422
	1.00	0.16682	0.167
0.70	0.40	0.88394	0.884
	0.70	0.71883	0.719
	1.00	0.46392	0.464
1.00	0.40	1.08684	1.087
	0.70	0.92243	0.922
	1.00	0.66712	0.667

Table 9.10.3. Transverse stresses distribution in the hollow cylinder of Figure 9.10.1

Coordinates		Transverse stresses	
x_3	R	SIOM	Exact
0.00	0.40	-0.44332	-0.444
	0.70	-0.44321	
	1.00	-0.44320	
0.25	0.40	-0.36009	-0.361
	0.70	-0.36121	
	1.00	-0.36104	
0.50	0.40	-0.11023	-0.111
	0.70	-0.11145	
	1.00	-0.11123	
0.70	0.40	0.30523	0.305
	0.70	0.30512	
	1.00	0.30514	
1.00	0.40	0.88872	0.889
	0.70	0.88912	
	1.00	0.88945	

Finally, as it can be seen from Tables 9.10.1 - 9.10.3 the numerical results by using the SIOM coincide very well with the exact solutions given in [44].

REFERENCES

[1] J.L. Swedlow and T.A. Cruse, Formulation of boundary integral equation for three-dimensional elastoplastic body, *Int. J. Solids Struct.* 7, 1673-1683 (1971).

[2] S. Mukherjee, Corrected boundary integral equation in planar thermoelastoplasticity, *Int. J. Solids Struct.* 13, 331-335 (1977).

[3] A. Mendelson, Boundary-integral methods in elasticity and plasticity, *NASA TN D 7418* (1973).

[4] H.D. Bui, Some remarks about the formulation of three-dimensional thermoelastoplastic problems of integral equations, *Int. J. Solids Struct.* 14, 935-939 (1978).

[5] J.C.F. Telles and C.A. Brebbia, On the application of the boundary element method to plasticity, *Appl. Math. Modelling* 3, 466-470 (1979).

[6] H. Poon, S. Mukherjee and M.F. Ahmad, Use of "simple solutions" in regularizing hypersingular boundary integral equations in elastoplasticity, *ASME J. Appl. Mech.* 65, 39-45 (1998).

[7] E.G. Ladopoulos, Singular integral representation of three-dimensional plasticity fracture problem, *Theor. Appl. Fract. Mech.* 8, 205-211 (1987).

[8] E.G. Ladopoulos, On the numerical evaluation of the singular integral equations used in two- and three-dimensional plasticity problems, *Mech. Res. Commun.* 14, 263-274 (1987).

[9] E.G. Ladopoulos, Singular integral operators method for two-dimensional plasticity problems, *Comp. Struct.* 33, 859-865 (1989).

[10] E.G. Ladopoulos, Singular integral operators method for two-dimensional elasto-plastic stress analysis, *Forsch. Ingen.* 57, 152-158 (1991).

[11] E.G. Ladopoulos, V.A. Zisis and D. Kravvaritis, Multidimensional singular integral equations in Lp applied to three-dimensional thermoelastoplastic stress analysis, *Comp. Struct.* 52, 781-788 (1994).

[12] E.G. Ladopoulos, *Singular Integral Equations, Linear and Non-linear Theory and its Applications in Science and Engineering,* Springer-Verlag, Berlin, New York (2000).

[13] O.C. Zienkiewicz, S. Valliappan and I.P. King, Elastoplastic solutions of engineering problems: initial stress finite element approach, *Int. J. Numer. Meth. Engng* 1, 75-100 (1969).

[14] G.C. Nayak and O.C. Zienkiewicz, Elastoplastic stress analysis. A generalization for various constitute relations including strain softening, *Int. J. Numer. Meth. Engng* 5, 113-135 (1972).

[15] W.F. Chen, *Limit Analysis and Soil Plasticity*, Elsevier, Amsterdam (1975).

[16] Y. Yamada, N. Yoshimura and T. Sakurai, Plastic stress-strain matrix and its application for the solution of elasto-plastic problems by the finite element method, *Int.J. Mech.Sci.*10, 343-354 (1968).

[17] P.V. Marcal and I.P. King, Elastic-plastic analysis of two-dimensional stress systems by the finite element method, *Int. J. Mech. Sci.* 9, 143-155 (1967).

[18] R. Hill, *The Mathematical Theory of Plasticity*, Oxford University Press, Oxford (1950).

[19] O. Hoffman and G. Sachs, Introduction to the Theory of Plasticity for Engineers, McGraw - Hill, New York (1953).
[20] P.G. Hodge Jr. and G.N. White Jr., A quantitative comparison of flow and deformation theories of plasticity, *J. Appl. Mech.* 17, 180-184 (1950).
[21] W.T. Koiter, Stress-strain relations, uniqueness and variational theorems for elastic-plastic materials with singular yield surface, *Q. Appl. Math.* 11, 350-354 (1953).
[22] D.C. Drucker and W. Pruger, Soil mechanics and plastic analysis at limit design, *Q. Appl. Mech.* 10, 157-165 (1952).
[23] W. Prager, The theory of plasticity - a survey of recent achievements. James Clayton lecture, *Proc. Inst. Mech. Engrs* 169, 41-57 (1955).
[24] D.N. De G. Allen and R.V. Southwell, Relaxation methods applied to engineering problems. XIV. Plastic straining in two-dimensional stress-systems, *Proc. R. Soc.* 242A, 379-414 (1950).
[25] W. Johnson and P.B. Mellor, *Plasticity for Mechanical Engineers,* Van Nostrand, London (1962).
[26] W.D. Liam Finn, Applications of limit plasticity in soil mechanics, *ASCE J. Soil Mech. Fdns Div.* SM 5, Part 1, 101-120 (1967).
[27] Z. Mroz, Non associated laws in plasticity, *J. Mech. Phys. Atmos.* 2, 21-41 (1963).
[28] J.N. Goodier and P.G. Hodge Jr., *The Mathematical Theory of Elasticity and Plasticity*, Wiley, New York (1958).
[29] J.H. Argyris, Elasto-plastic matrix displacement analysis of three-dimensional continua, *J. R.Aeronaut. Soc.* 69, 633-635 (1965).
[30] G.G. Pope, *A Discrete Element Method for Analysis of Plane Elasto-plastic Strain Problems*, R.A.F. Farnborough T.R. 65028 (1965).
[31] R.von Mises, Mechanik der plastischen formanderung der kristallen, *ZAMM* 8, 161-185 (1928).
[32] J. Casey and P.M. Naghdi, On the characterization of strain-hardening in plasticity, *ASME J. Appl.Mech.* 48, 285-295 (1981).
[33] J. Casey and P.M. Naghdi, On the nonequivalence of the stress space and strain space formulation of plasticity theory, *ASME J. Appl. Mech.* 50, 350-354 (1983).
[34] J. Casey and P.M. Naghdi, A remark on the definition of hardening, softening and perfectly plastic behaviour, *Acta Mech.* 48, 91-94 (1983).
[35] P.M. Naghdi and J.A. Trapp, The significance of formulating plasticity theory with reference to loading surfaces in strain space, *Int. J. Engng Sci.* 13, 785-797 (1975).
[36] P.V. Lade, Elasto-plastic stress-strain theory for cohesionless soil with curved yield surfaces, *Int. J.Solids Struct.* 13, 1019-1035 (1977).
[37] P.J. Yoder and W.D. Irwan, On the formulation of strain-space plasticity with multiple loading surfaces, *ASME J. Appl. Mech.* 48, 773-778 (1981).
[38] J.H. Prevost, Plasticity theory for soils stress-strain behavior, *ASCE J. Engng Mech.* 104, 1177-1196 (1978).
[39] P.K. Banerjee and A.S. Stipho, An elastoplastic model for undrained behavior of heavily overconsolidated clay, *Int. J. Numer. Anal. Meth. Geom.* 3, 97-103 (1979).
[40] P.S. Theocaris and E. Marketos, Elastic-plastic analysis of perforated thin strips of strain-hardening material, *J. Mech. Phys. Solids* 12, 377-390 (1964).
[41] S.F. Reyes and D.U. Deere, Elastic-plastic analysis of underground openings by the finite element method, *Proc. 1st Congr. Rock Mechanics,* 477-483, Lisbon (1966).

[42] A.C.T. Chen and W.F. Chen, Constitutive equations and punch-indentation of concrete, *Proc. ASCE J. Engng Mech. Div.* 101, 889-906 (1975).
[43] F. John, *Plane Waves and Spherical Means Applied to Partial Differential Equations*, Interscience, New York (1955).
[44] B.A. Boley and J.H. Weiner, *Theory of Thermal Stresses,* J.Wiley, New York (1960).

APPENDIX

MATHEMATICAL DEFINITIONS

Following definitions are mentioned, as are used often in this book.

Definition A1

Consider by D a closed or open domain of the real numbers R. By contour L in C, is defined a continuous image L of the interval $D \subseteq R$, i.e., a subtotal $L \subseteq R$ is called contour, if and only if, there exists a domain $D \subseteq R$ and a continuous function $z = z(t)$, $\forall t \in D$, such as to be valid :

$$L = \{ z \in C : z = z(t), \forall t \in D \} \tag{A1.1}$$

Definition A2

Consider the contour L, defined by relation (A1.1). If $z = z(t)$ is an equivalence relation of domain D in C, i.e., if there exists $z(t) \neq z(t')$, $\forall t \neq t'$, with $t, t' \in D$, then the contour L is called to be simple.

Definition A3

Consider the contour L, defined by relation (A1.1). If for at least one representation of L it is valid : $\exists z'(t) \neq 0$, $\forall t \in D$ and is continuous in D, then the contour is called to be smooth.

Definition A4

Consider the contour L, defined by relation (A1.1). This contour is called to be closed, if and only if, its edges coincide each other, i.e., if there exists $z(a + 0) = z(b - 0)$, where a and b are respectively the left and the right edge of the domain D in $[-\infty, +\infty]$.

Definition A5

Lyapounov contour is called to be every simple, smooth and closed contour in which the angle θ between the tangents in two points t_1 and t_2 of the contour satisfies the inequality :

$$\theta \leq A \mid t_1 - t_2 \mid^{\mu} \tag{A5.1}$$

where A and μ are positive constants.

Definition A6

The complex function $f(t)$ is said to satisfy on the smooth curve L, the Hölder condition (H-condition), if for two arbitrary points of the curve it is valid :

$$\mid f(t_2) - f(t_1) \mid < A \mid t_2 - t_1 \mid^{\mu} \tag{A6.1}$$

where $A > 0$ and $0 < \mu \leq 1$. A is called the Hölder constant and μ the Hölder index.

Definition A7

If the complex function $f(t,x)$ has two variables, then is said to satisfy on the smooth curve L, the Hölder condition (H-condition), if for any pair values (t_1, t_2), (x_1, x_2) belonging to the curve, it is valid :

$$\mid f(t_2, x_2) - f(t_1, x_1) \mid < A \mid t_2 - t_1 \mid^{\mu} + B \mid x_2 - x_1 \mid^{\nu} \tag{A7.1}$$

where A, B, μ, ν are positive numbers, and μ, $\nu \leq 1$.

Definition A8

First order Fredholm equation, is said to be every integral equation of the form:

$$\int_L K(x,t)\varphi(x)\,\mathrm{d}x = f(t) \tag{A8.1}$$

where x, t are points of the integration contour L, $K(x,t)$ and $f(t)$ known functions, $\varphi(x)$ the unknown function and following conditions are valid :

$$\int_L \int_L \mid K(x,t) \mid^2 \mathrm{d}x\,\mathrm{d}t < \infty \tag{A8.2}$$

and: $$\int_L |f(t)|^2\,\mathrm{d}t < \infty \tag{A8.3}$$

Definition A9

Second order Fredholm equation, is said to be every integral equation of the form :

$$\varphi(t) - \lambda \int_L K(x,t)\varphi(x)\,\mathrm{d}x = f(t) \tag{A9.1}$$

where $K(x,t)$, $f(t)$ are known functions, $\varphi(x)$ is the unknown function, x, t are points of the integration contour L, λ an arbitrary complex number and conditions (A8.2) and (A8.3) are valid.

Definition A10

The linear operator $L: X \rightarrow Y$, where X, Y Banach spaces, is said to be compact, if the $L(B_x)$ is included in a compact subset of Y, where :

$$B_x = \{ x \in X : \|x\| \leq 1 \} \tag{A10.1}$$

Definition A11

A set A in a linear space X is said to be convex, if for any pair $(x,y) \in A$, points $\{ tx + (1 - t)y \}$, with $0 < t < 1$, are also included in A.

Definition A12

By diameter of a set A , is called a metric space (X, d) according to the relation :

$$\delta(A) = \sup_{\substack{x \in A \\ y \in A}} d(x, y) \tag{A12.1}$$

SUBJECT INDEX

A

B

C

D

E

F

G

H

I

S

T

U

V

W

Y

Author Index

A

B

C

D

E

F

N

O

P

R

S

T

V

W

Y

Z